Advances in MATHEMATICAL ECONOMICS

Aims and Scope. The project is to publish *Advances in Mathematical Economics* once a year under the auspices of the Research Center of Mathematical Economics. It is designed to bring together those mathematicians who are seriously interested in obtaining new challenging stimuli from economic theories and those economists who are seeking effective mathematical tools for their research.

The scope of *Advances in Mathematical Economics* includes, but is not limited to, the following fields:

- Economic theories in various fields based on rigorous mathematical reasoning.
- Mathematical methods (e.g., analysis, algebra, geometry, probability) motivated by economic theories.
- Mathematical results of potential relevance to economic theory.
- Historical study of mathematical economics.

Authors are asked to develop their original results as fully as possible and also to give a clear-cut expository overview of the problem under discussion. Consequently, we will also invite articles which might be considered too long for publication in journals.

Springer
Tokyo
Berlin
Heidelberg
New York
Hong Kong
London
Milan
Palis

S. Kusuoka, T. Maruyama (Eds.)

Advances in Mathematical Economics

Volume 6

Springer

Shigeo Kusuoka
Professor
Graduate School of Mathematical Sciences
University of Tokyo
3-8-1 Komaba, Meguro-ku
Tokyo, 153-0041 Japan

Toru Maruyama
Professor
Department of Economics
Keio University
2-15-45 Mita, Minato-ku
Tokyo, 108-8345 Japan

ISBN 4-431-20315-X Springer-Verlag Tokyo Berlin Heidelberg New York

Printed on acid-free paper
Springer-Verlag is a company in the BertelsmannSpringer publishing group.

Printed in Japan

Camera-ready copy prepared from the authors' LaTeXfiles.
Printed and bound by Hirakawa Kogyosha, Japan.
SPIN: 10966679

Table of Contents

Research Articles

Adv. Math. Econ. 6, 1–38 (2004)

Advances in
MATHEMATICAL ECONOMICS

On the fiber product of Young measures with application to a control problem with measures

Charles Castaing[1] **and Paul Raynaud de Fitte**[2]

[1] Département de Mathématiques, Case 051, Université de Montpellier II, 34095 Montpellier Cedex 5, France
[2] Laboratoire Raphaël Salem, UMR CNRS 6085, UFR Sciences, Université de Rouen, 76821 Mont Saint Aignan Cedex, France

Received: April 11, 2003
Revised: June 19, 2003

JEL classification: C61, C73

Mathematics Subject Classification (2000): 28 A 33, 46 N 10, 49 L 25

Abstract. This paper studies, in the context of separable metric spaces, the stable convergence of the fiber product for Young measures with applications to a control problem governed by an ordinary differential equations where the controls are Young measures. Essentially we study some variational properties of the value functions and the existence of quasi-saddle points of these functions which occurs in this dynamic control problem, and also their link with the viscosity solution of the associated Hamilton–Jacobi–Bellman equation.

Key words: Young measure, relaxed control, fiber product, dynamic programming, viscosity solution

1. Introduction

This paper is divided in two parts. In section 2 we state some new convergence results for Young measures and also the proofs of the fiber product of Young measures. The third section is devoted to the study of the value functions of a control problem where the dynamic is governed by an ordinary differential equation (ODE) where the controls are Young measures. Here the stable convergence for the fiber product of Young measures is crucial in the statement of the variational properties of the value functions in the control problems under consideration and the developements of Mathematical Economics (see e.g [Tat02]). Similar differential games with Young measures governed by some classes of evolution inclusions are given in a forthcoming paper. References for control problems are e.g [EK72 Ell87 ES84 KS88 BJ91].

2. Stable convergence versus convergence in probability

2.1 Young measures

For simplicity, the topological spaces we consider in this work are only metric spaces. Most of the convergence results on Young measures in this work can be extended to the case of completely regular Suslin spaces (which includes e.g. weak topologies of separable Banach spaces, or spaces of distributions). This will be detailed in a forthcoming work. All metric spaces we shall consider are assumed to be separable, or, more generally, they do not contain any discrete subset with measurable cardinal: this ensures that every Borel measure μ on such a metric space $\mathbb{S}$ has a separable support and, consequently, μ is inner regular w.r.t. the totally bounded subsets of $\mathbb{S}$ (the reader interested in measurable cardinals can read [Bil68, Appendix III]; recall that it is consistent with the usual axioms of logic to assume that measurable cardinals do not exist). Actually, in the convergence results presented here, we only need that the limit has separable support.

If $(\mathbb{S}, d)$ is a metric space, we denote by $C_b(\mathbb{S})$ the set of all real-valued bounded continuous functions defined on $\mathbb{S}$ and we denote by $\mathcal{B}_{\mathbb{S}}$ the Borel σ–algebra of $\mathbb{S}$. The set of probability measures on $\mathcal{B}_{\mathbb{S}}$ is denoted by $\mathcal{M}^1_+(\mathbb{S})$. We endow $\mathcal{M}^1_+(\mathbb{S})$ with the narrow (or weak) topology, that is, the coarsest topology such that, for each $f \in C_b(\mathbb{S})$, the mapping $\mu \mapsto \mu(f)$, $\mathcal{M}^1_+(\mathbb{S}) \to \mathbb{R}$, is continuous.

Let $\mathrm{BL}(\mathbb{S}, d) = \mathrm{BL}(d)$ be the set of all mappings $f : \mathbb{S} \to \mathbb{R}$ which satisfy

$$\|f\|_{\mathrm{BL}(d)} := \sup_{x\in\mathbb{S}} |f(x)| + \sup_{x,y\in\mathbb{S},\, x\neq y} \frac{|f(x)-f(y)|}{d(x,y)} < +\infty.$$

The space $\mathrm{BL}(d)$, endowed with the norm $\|.\|_{\mathrm{BL}(d)}$, is a Banach space. Dudley [Dud66, Theorem 6 and Theorem 8] has shown that $\mathcal{M}^1_+(\mathbb{S})$ embeds homeomorphically in the strong dual $\mathrm{BL}(d)^*$ of $\mathrm{BL}(d)$. Let us denote by $\mathrm{BL}_1(d)$ the unit ball of $\mathrm{BL}(d)$. The topology of $\mathcal{M}^1_+(\mathbb{S})$ is induced by the metric $D_{\mathrm{BL}(d)}$ defined by

$$D_{\mathrm{BL}(d)}(\mu, \nu) = \sup_{f\in\mathrm{BL}_1(d)} (\mu(f) - \nu(f)).$$

Let $(\mathbb{S}, d)$ be a metric space and let $(\Omega, \mathcal{S}, \mathrm{P})$ be a probability space. We denote by $\mathcal{Y}(\Omega, \mathcal{S}, \mathrm{P}; \mathbb{S})$ the set of measurable mappings

$$\mu : \begin{cases} \Omega & \mapsto & \mathcal{M}^1_+(\mathbb{S}) \\ \omega & \mapsto & \mu_\omega \end{cases}$$

Each element μ of $\mathcal{Y}(\Omega, \mathcal{S}, \mathrm{P}; \mathbb{S})$ can be identified with the measure $\widetilde{\mu}$ on $(\Omega \times \mathbb{S}, \mathcal{S} \otimes \mathcal{B}_{\mathbb{S}})$ defined by $\widetilde{\mu}(A \times B) = \int_A \mu_\omega(B)\, d\mathrm{P}(\omega)$ (and the mapping $\mu \mapsto \widetilde{\mu}$ is onto if $\mathbb{S}$ has the Radon property, see e.g. [Val73]). In the sequel, we shall use freely this identification. For instance, if $f : \Omega \times \mathbb{S} \to \mathbb{R}$ is a bounded measurable mapping, the notation $\mu(f)$ denotes $\int_\Omega \mu_\omega(f(\omega, .))\, d\mathrm{P}(\omega)$. The elements of $\mathcal{Y}(\Omega, \mathcal{S}, \mathrm{P}; \mathbb{S})$ are called *Young measures* on $\Omega \times \mathbb{S}$.

The set $\mathcal{Y}(\Omega, \mathcal{S}, \mathrm{P}; \mathbb{S})$ is endowed with the coarsest topology such that, for each $A \in \mathcal{S}$ and each $f \in C_b(\mathbb{S})$, the mapping $\mu \mapsto \mu(\mathbf{1}_A \otimes f)$, $\mathcal{Y}(\Omega, \mathcal{S}, \mathrm{P}; \mathbb{S}) \to \mathbb{R}$ is continuous. Convergence in this topology is sometimes called *stable convergence*, we will follow this tradition.

Let $\mathrm{L}^0(\Omega; \mathbb{S})$ be the set of random elements of $\mathbb{S}$ defined on Ω (we identify random elements which are equal P–almost everywhere). To each element X of $\mathrm{L}^0(\Omega; \mathbb{S})$, we associate the Young measure $\underline{\delta}_X : \omega \mapsto \delta_{X(\omega)}$, where, for any $x \in \mathbb{S}$, δ_x denotes the probability concentrated on x. Young measures of the form $\underline{\delta}_X$ are called *degenerate Young measures*. If P has no atoms and if $\mathbb{S}$ is Suslin, the set of degenerate Young measures is dense in $\mathcal{Y}(\Omega, \mathcal{S}, \mathrm{P}; \mathbb{S})$, see [Bal84b].

We call *integrand* on $\Omega \times \mathbb{S}$ any measurable mapping $f : \Omega \times \mathbb{S} \to \mathbb{R}$. If furthermore $f(\omega, .)$ is continuous for every $\omega \in \Omega$, we say that f is a *Carathéodory integrand*. In the case when $\mathbb{S}$ is separable, a sufficient condition for a mapping $f : \Omega \times \mathbb{S} \to \mathbb{R}$ to be a Carathéodory integrand is that $f(\omega, .)$ be continuous for every $\omega \in \Omega$ and $f(., x)$ be measurable for every $x \in \mathbb{S}$ [CV77, Lemma III.14]. If f is an integrand such that there exists a measurable P–integrable function $\phi : \Omega \to \mathbb{R}^+$ such that $|f(\omega, .)| \leq \phi(\omega)$ for each $\omega \in \Omega$, we say that f is *L^1-bounded*.

It is well known that, if $\mathbb{S}$ is a Suslin metrizable space and if $(\mu^\alpha)_\alpha$ is a net in $\mathcal{Y}(\Omega, \mathcal{S}, \mathrm{P}; \mathbb{S})$ which stably converges to some $\mu^\infty \in \mathcal{Y}(\Omega, \mathcal{S}, \mathrm{P}; \mathbb{S})$, then we have $\liminf_\alpha \mu^\alpha(f) \geq \mu^\infty(f)$ for every integrand $f \geq 0$ such that $f(\omega, .)$ is l.s.c. for each $\omega \in \Omega$ (this is called the Portmanteau Theorem, or the Semicontinuity Theorem, see e.g. [Bal00] for a proof for sequences – the result also holds true for nets, see [RdF03] for a reasoning with nets). In particular, we have the following characterization of stable convergence.

Stable convergence and Carathéodory integrands *Assume that $\mathbb{S}$ is Suslin metrizable, let $(\mu^\alpha)_\alpha$ be a net in $\mathcal{Y}(\Omega, \mathcal{S}, \mathrm{P}; \mathbb{S})$ and let $\mu^\infty \in \mathcal{Y}(\Omega, \mathcal{S}, \mathrm{P}; \mathbb{S})$. Then $(\mu^\alpha)_\alpha$ stably converges to μ^∞ if and only if* $\lim_\alpha \mu^\alpha(f) = \mu^\infty(f)$ *for every L^1-bounded Carathéodory integrand f on $\Omega \times \mathbb{S}$.*

In this section, we shall also use a different result, which holds for nonnecessarily Suslin spaces. Let us first define the spaces $\underline{\mathrm{BL}}(d)$ and $\underline{\mathrm{BL}}'(d)$.

We denote by $\underline{\mathrm{BL}}(d)$ the set of all integrands f on $\Omega \times \mathbb{S} \to \mathbb{R}$ such that there exists a measurable mapping $\phi : \Omega \to \mathbb{R}^+$ with $\|f(\omega,.)\|_{\mathrm{BL}(d)} \leq \phi(\omega)$ for every $\omega \in \Omega$ (if $\mathbb{S}$ is separable, $\underline{\mathrm{BL}}(d)$ is simply the set of integrands f such that $f(\omega,.) \in \mathrm{BL}(d)$ for each $\omega \in \Omega$). We denote by $\underline{\mathrm{BL}}'(d)$ the set of elements f of $\underline{\mathrm{BL}}(d)$ which have the form

$$f(\omega,x) = \sum_{i=1}^{n} \mathbf{1}_{A_i}(\omega) f_i(x),$$

where $(A_1,\dots,A_n)$ is a measurable partition of Ω and each f_i is in $\mathrm{BL}(d)$. Let $L^1_{\mathrm{BL}(d)}$ be the space of Bochner integrable functions defined on $(\Omega, \mathcal{S}, \mathrm{P})$ with values in $\mathrm{BL}(d)$. We have

$$\underline{\mathrm{BL}}'(d) \subset L^1_{\mathrm{BL}(d)} \subset \left\{ f \in \underline{\mathrm{BL}}(d);\ f \text{ is } L^1\text{-bounded} \right\}.$$

Lemma 2.1.1. *Let $(\mathbb{S}, d)$ be a metric space. Let $(\mu^\alpha)_{\alpha\in\mathbb{A}}$ be a net in $\mathcal{Y}(\Omega, \mathcal{S}, \mathrm{P}; \mathbb{S})$ and let $\mu^\infty \in \mathcal{Y}(\Omega, \mathcal{S}, \mathrm{P}; \mathbb{S})$. The following conditions are equivalent.*

(a) *$(\mu^\alpha)_\alpha$ stably converges to μ^∞.*
(b) *For each L^1-bounded integrand $f \in \underline{\mathrm{BL}}(d)$, we have $\lim_\alpha \mu^\alpha(f) = \mu^\infty(f)$.*
(c) *For each $f \in L^1_{\mathrm{BL}(d)}$, we have $\lim_\alpha \mu^\alpha(f) = \mu^\infty(f)$.*
(d) *For each integrand $f \in \underline{\mathrm{BL}}'(d)$, we have $\lim_\alpha \mu^\alpha(f) = \mu^\infty(f)$.*

Proof. Assume (a). Let $\epsilon > 0$. Let f be an L^1-bounded element of $\underline{\mathrm{BL}}(d)$. There exists a measurable function $\phi : \Omega \to \mathbb{R}^+$ such that $\|f(\omega,.)\|_{\mathrm{BL}(d)} \leq \phi(\omega)$ for every $\omega \in \Omega$. Furthermore, there exists a P–integrable function $\varphi : \Omega \to \mathbb{R}^+$ such that $|f(\omega,x)| \leq \varphi(\omega)$ for each $(\omega,x) \in \Omega \times \mathbb{S}$. We can thus find $\Omega_\epsilon \in \mathcal{S}$ and $M > 0$ such that

$$\mathrm{P}(\Omega \setminus \Omega_\epsilon) < \epsilon, \quad \phi\,\mathbf{1}_{\Omega_\epsilon} \leq M, \quad \text{and} \quad \int_{\Omega\setminus\Omega_\epsilon} \varphi \, d\mathrm{P} < \epsilon.$$

Define $\widetilde{f} \in \underline{\mathrm{BL}}(d)$ by

$$\widetilde{f}(\omega,x) = \begin{cases} \frac{1}{M} f(\omega,x) & \text{if } \omega \in \Omega_\epsilon \\ 0 & \text{if } \omega \in \Omega \setminus \Omega_\epsilon \end{cases}$$

We have $\widetilde{f}(\omega,.) \in \mathrm{BL}_1(d)$ for every $\omega \in \Omega$. Moreover, for any $\mu \in \mathcal{Y}(\Omega, \mathcal{S}, \mathrm{P}; \mathbb{S})$, we have

$$\left|\mu(f - M\widetilde{f})\right| \leq \epsilon. \tag{2.1.1}$$

Now, from our general hypothesis on metric spaces (see the Introduction), the measure $\mu^\infty(\Omega \times .) \in \mathcal{M}^1_+(\mathbb{S})$ has a separable support, thus it is inner regular w.r.t. the totally bounded subsets of $\mathbb{S}$. There exists a totally bounded subset K of $\mathbb{S}$ such that $\mu^\infty(\Omega \times K) > 1 - \epsilon/M$. Recall that every Lipschitz function $h : K \to [0,1]$ can be extended to a Lipschitz function defined on $\mathbb{S}$, with same Lipschitz coefficient, and with values in $[0,1]$ (see e.g. [Dud66]). For any continuous function h on $\mathbb{S}$ and any $B \subset \mathbb{S}$, let us denote $\|h\|_B := \sup_{x\in B} |h(x)|$. The set of restrictions to K of elements of $\mathrm{BL}_1(d)$ is totally bounded for $\|.\|_K$ (it is a subset of the compact space $\mathrm{BL}_1(\widehat{K}, d)$, where $\widehat{K}$ is the d–completion of K). e There exist thus $g_1, \dots, g_n \in \mathrm{BL}_1(d)$ such that, for each $h \in \mathrm{BL}_1(d)$, we have $\inf_{i=1,\dots,n} \|h - g_i\|_K \le \epsilon/M$.

For each $\omega \in \Omega$, we can find $N(\omega) \in \{1, \dots, n\}$ such that $\|\widetilde{f}(\omega, .) - g_{N(\omega)}\|_K \le \epsilon/M$. Furthermore, we can assume that N is measurable, because Lipschitz functions on K are determined by their values on a countable (dense) subset of K. For $i = 1, \dots, n$, let $A_i = \{N(\omega) = i\}$, and let $g = \sum_{i=1}^n \mathbf{1}_{A_i} \otimes g_i$. We have $g \in \underline{\mathrm{BL}}'(d)$ and $\|\widetilde{f}(\omega, .) - g(\omega, .)\|_K \le \epsilon/M$ for every $\omega \in \Omega$.

Let $K^\epsilon = \{x \in \mathbb{S};\ d(x, K) < \epsilon/M\}$. For each $\omega \in \Omega$, as $\widetilde{f}(\omega, .)$ and $g(\omega, .)$ are 1–Lipschitz, we have

$$\left\|\widetilde{f}(\omega, .) - g(\omega, .)\right\|_{K^\epsilon} \le 3\frac{\epsilon}{M}. \tag{2.1.2}$$

Let $h : \mathbb{S} \to [0,1]$ be a Lipschitz mapping such that $h(x) = 1$ if $x \in K$ and $h(x) = 0$ if $x \notin K^\epsilon$ (we can take e.g. $h(x) = (1 - (M/\epsilon)d(x, K)) \vee 0$). We have

$$\lim_\alpha \mu^\alpha(\mathbf{1}_\Omega \otimes (1 - h)) = \mu^\infty(\mathbf{1}_\Omega \otimes (1 - h)) \le \mu^\infty(\Omega \times (\mathbb{S} \setminus K)) \le \epsilon/M,$$

thus there exists $\alpha_0 \in \mathbb{A}$ such that

$$\alpha \ge \alpha_0 \Rightarrow \mu^\alpha(\mathbf{1}_\Omega \otimes (1 - h)) \le 2\frac{\epsilon}{M}. \tag{2.1.3}$$

Furthermore, we have $\left|\widetilde{f}(\omega, x) - g(\omega, x)\right| \le 2$ for every $(\omega, x) \in \Omega \times \mathbb{S}$, because $\widetilde{f}$ and g are bounded by 1. From (2.1.2) and (2.1.3) we thus have, for $\alpha \ge \alpha_0$,

$$\begin{aligned}
\left|(\mu^\alpha - \mu^\infty)\left(\widetilde{f} - g\right)\right| &\le (\mu^\alpha + \mu^\infty)\left|\widetilde{f} - g\right| \\
&\le (\mu^\alpha + \mu^\infty)\left(\left|\widetilde{f} - g\right|(\mathbf{1}_\Omega \otimes h)\right) \\
&\qquad + (\mu^\alpha + \mu^\infty)\left(2(\mathbf{1}_\Omega \otimes (1 - h))\right) \\
&\le 3\frac{\epsilon}{M} + 3\frac{\epsilon}{M} + 4\frac{\epsilon}{M} + 2\frac{\epsilon}{M} = 12\frac{\epsilon}{M}.
\end{aligned}$$

But we also have $\lim_\alpha (\mu^\alpha - \mu^\infty)(g) = 0$, thus, taking α_0 large enough, we have, for $\alpha \geq \alpha_0$,

$$\left|(\mu^\alpha - \mu^\infty)\left(\widetilde{f}\right)\right| \leq 13\frac{\epsilon}{M},$$

and thus, using (2.1.1),

$$|(\mu^\alpha - \mu^\infty)(f)| \leq \left|(\mu^\alpha - \mu^\infty)\left(f - M\widetilde{f}\right)\right| + M\left|(\mu^\alpha - \mu^\infty)\left(\widetilde{f}\right)\right| \leq 15\epsilon$$

which yields *(b)* because ϵ is arbitrary.

The implications *(b)⇒(c)⇒(d)* are clear. Assume now *(d)*. We have $\lim_\alpha \mu^\alpha(\mathbf{1}_A \otimes g) = \mu^\infty(\mathbf{1}_A \otimes g)$ for each $A \in \mathcal{S}$ and each $g \in \mathrm{BL}_1(d)$. From [Dud66, Theorem 8], this means that, for each $A \in \mathcal{S}$, the sequence of measures $\mu^\alpha(A \times .)$ narrowly converges to $\mu^\infty(A \times .)$. We thus have, for each $A \in \mathcal{S}$ and for each $g \in \mathcal{C}_b(\mathbb{S})$, $\lim_\alpha \mu^\alpha(\mathbf{1}_A \otimes g) = \mu^\infty(\mathbf{1}_A \otimes g)$, which proves *(a)*. □

From the Dudley embedding theorem [Dud66], each Young measure $\mu \in \mathcal{Y}(\Omega, \mathcal{S}, \mathrm{P}; \mathbb{S})$ can be viewed as an element of $L^\infty_{\mathrm{BL}(\mathbb{S})^*}$, where $\mathrm{BL}(d)^*$ denotes the topological dual of $\mathrm{BL}(d)$. We thus have the following immediate consequence of Lemma 2.1.1.

Corollary 2.1.2. *Let $(\mathbb{S}, d)$ be a metric space. On $\mathcal{Y}(\Omega, \mathcal{S}, \mathrm{P}; \mathbb{S})$, the stable topology coincides with $\sigma(L^\infty_{\mathrm{BL}(d)^*}, L^1_{\mathrm{BL}(d)})$.*

Remark. Similarly, from the characterization of stable convergence with the Carathéodory integrands, it is not difficult to show that, *if $\mathbb{S}$ is a Suslin metrizable space, the stable topology on $\mathcal{Y}(\Omega, \mathcal{S}, \mathrm{P}; \mathbb{S})$ coincides with $\sigma(L^\infty_{\mathcal{C}_b(\mathbb{S})^*}, L^1_{\mathcal{C}_b(\mathbb{S})})$*. Actually, the proof is simpler than that of Lemma 2.1.1, because the Lipschitz coefficient does not need to be controlled.

Compactness and convex combinations

We now gather some results which we shall use later, in order to simplify some hypothesis. In the reminder of this subsection, $\mathbb{S}$ will be a Polish space. A subset $\mathcal{H}$ of $\mathcal{Y}(\Omega, \mathcal{S}, \mathrm{P}; \mathbb{S})$ is said to be *tight* if, for each $\epsilon > 0$, there exists a graph measurable random compact subset K of $\mathbb{S}$ such that $\inf_{\mu \in \mathcal{H}} \int_\Omega \mu_\omega(K(\omega)) \geq 1 - \epsilon$. From a well known result (Valadier in [Jaw84], Balder [Bal84a]), $\mathcal{H}$ is tight if and only if, for each $\epsilon > 0$, there exists a compact subset K of $\mathbb{S}$ such that $\inf_{\mu \in \mathcal{H}} \mu(\Omega \times K) \geq 1 - \epsilon$. In other words, $\mathcal{H}$ is tight if and only if the subset $\{\mu(\Omega \times .); \mu \in \mathcal{H}\}$ of $\mathcal{M}^1_+(\mathbb{S})$ is tight in the usual sense. It is well known that $\mathcal{H}$ is relatively compact (or, equivalently, relatively sequentially compact) for the stable topology if and only if it is tight: see [Bal89 Bal90 JM81 JM83 RdF03].

Let us recall a useful result of Balder.

K–convergence [Bal89 Bal90 Bal00] *Let $(\mathbb{S}, d)$ be a Polish space. Let $(\mu^n)_n$ be a sequence in $\mathcal{Y}(\Omega, \mathcal{S}, \mathrm{P}; \mathbb{S})$ which stably converges to some $\mu \in \mathcal{Y}(\Omega, \mathcal{S}, \mathrm{P}; \mathbb{S})$. Then each subsequence of $(\mu^n)_n$ contains a further subsequence $(\widetilde{\mu}^n)_n$ which K–converges to μ, that is, for each subsequence $(\nu^n)_n$ of $(\widetilde{\mu}^n)_n$, we have*

$$\lim_n \frac{1}{n} \sum_{i=1}^{n} \nu_\omega^i = \mu_\omega \text{ a.e.}$$

Proposition 2.1.3. *Let $\mathbb{S}$ be a Polish space. Let $\mathcal{H} \subset \mathcal{Y}(\Omega, \mathcal{S}, \mathrm{P}; \mathbb{S})$. The following conditions are equivalent.*

(i) *$\mathcal{H}$ is relatively compact for the stable topology.*

(ii) *$\mathcal{H}$ has the* Mazur property, *that is, for every sequence (λ^n) in $\mathcal{H}$, there exists a sequence $(\mu^n)_n$ with $\mu^n \in \mathrm{co}(\{\lambda^m : m \geq n\})$ such that, for P–almost every ω, $(\mu_\omega^n)_n$ converges narrowly in $\mathcal{M}_+^1(\mathbb{S})$.*

Proof. Assume (i). We know that relative compactness for the stable topology amounts to relative sequential compactness. From Balder's result on K–convergence, for each sequence $(\mu^n)_n$ of elements of $\mathcal{H}$, there exists a subsequence $(\lambda^n)_n$ which K–converges to some $\lambda \in \mathcal{Y}(\Omega, \mathcal{S}, \mathrm{P}; \mathbb{S})$. For almost every $\omega \in \Omega$, the sequence

$$(\nu_\omega^n)_n := \left(\frac{1}{n(n-1)} \sum_{i=n+1}^{n^2} \lambda_\omega^i \right)_n = \left(\frac{n^2}{n^2 - n} \left(\frac{1}{n^2} \sum_{i=1}^{n^2} \lambda_\omega^i \right) \right)_n - \left(\frac{1}{n-1} \left(\frac{1}{n} \sum_{i=1}^{n} \lambda_\omega^i \right) \right)_n$$

is a.e. convergent, and we have $\nu^n \in \mathrm{co}\,\{\mu^m;\, m \geq n\}$ for every $n \geq 1$, which proves (ii).

Assume (ii). Let $(\mu^n)_n$ be a sequence of elements of $\mathcal{H}$. For each subsequence $(\lambda^n)_n$, there exists a sequence $(\nu^n)_n$ in $\mathcal{Y}(\Omega, \mathcal{S}, \mathrm{P}; \mathbb{S})$ such that $\nu^n \in \mathrm{co}\,\{\lambda^m;\, m \geq n\}$ for every $n \in \mathbb{N}$ and such that $(\nu_\omega^n)_n$ a.e. converges. In particular, $(\nu^n(\Omega \times .))_n$ is narrowly convergent thus, as $\mathbb{S}$ is Polish, (ν^n) is tight in the sense of Young measures, that is, the sequence of margins $(\nu^n(\Omega \times .))_n$ is tight in the usual sense. Thus, for each subsequence $(\lambda^n(\Omega \times .))_n$ of $(\mu^n(\Omega \times .))_n$, there exists a narrowly convergent sequence $(\widetilde{\nu}^n)_n$ in $\mathcal{M}_+^1(\mathbb{S})$ such that $\widetilde{\nu}^n \in \mathrm{co}\,\{\lambda^m(\Omega \times .);\, m \geq n\}$ for every $n \in \mathbb{N}$. From [CV98, Proposition 3.3], this implies that $(\mu^n(\Omega \times .))_n$ is tight, that is, $(\mu^n)_n$ is tight in the sense of Young measures. This proves (i). □

2.2 Convergence in probability of Young measures

Recall that a sequence (X_n) of random elements of a metric space $(\mathbb{S}, d)$ converges in probability to a random element X if one of the following equivalent conditions is satisfied.

$$\lim_n \mathrm{P}(\{d(X_n - X) \geq \epsilon\}) = 0, \ \forall \epsilon > 0$$

$$\lim_n \mathrm{E}(d(X_n - X) \wedge 1) = 0.$$

Note that this definition is topological: it does not depend on the metric d, but only on the topology of $(\mathbb{S}, d)$ (see [HJ91 HJ98]). The topology of convergence in probability on $\mathrm{L}^0(\Omega; \mathbb{S})$ is induced by the metric Δ_d defined by

$$\Delta_d(X, Y) = \int_\Omega d(X, Y) \wedge 1 \, d\mathrm{P}.$$

Thus, on the space $\mathcal{Y}(\Omega, \mathcal{S}, \mathrm{P}; \mathbb{S}) = \mathrm{L}^0\left(\Omega; \mathcal{M}^1_+(\mathbb{S})\right)$, the topology of convergence in probability is induced by the metric $\Delta_{\underline{\mathrm{BL}}(d)} := \Delta_{(D_{\mathrm{BL}(d)})}$ defined by

$$\Delta_{\underline{\mathrm{BL}}(d)}(\mu, \nu) = \int_\Omega D_{\mathrm{BL}(d)}(\mu_\omega, \nu_\omega) \, d\mathrm{P}(\omega).$$

Let us define this metric in a more convenient way. Let

$$\begin{aligned} &\underline{\mathrm{BL}}_1(d) = \{f \in \underline{\mathrm{BL}}(d);\ f(\omega, .) \in \mathrm{BL}_1(d) \text{ a.e.}\} \\ \text{and}\quad &\underline{\mathrm{BL}}'_1(d) = \left\{f \in \underline{\mathrm{BL}}'(d)\,;\ f(\omega, .) \in \mathrm{BL}_1(d) \text{ a.e.}\right\}. \end{aligned}$$

Theorem 2.2.1. *Let $(\mathbb{S}, d)$ be a metric space. The topology of convergence in probability on $\mathcal{Y}(\Omega, \mathcal{S}, \mathrm{P}; \mathbb{S})$ is induced by $\Delta_{\underline{\mathrm{BL}}(d)}$ and we have, for all $\mu, \nu \in \mathcal{Y}$,*

$$\Delta_{\underline{\mathrm{BL}}(d)}(\mu, \nu) = \sup_{f \in \underline{\mathrm{BL}}_1(d)} (\mu(f) - \nu(f)) = \sup_{f \in \underline{\mathrm{BL}}'_1(d)} (\mu(f) - \nu(f)). \tag{2.2.1}$$

Proof. The first part of Theorem 2.2.1 is clear from the discussion we had on convergence in probability. There remains to prove the equalities (2.2.1). Let $\mu, \nu \in \mathcal{Y}$. We have

$$\begin{aligned} \sup_{f \in \underline{\mathrm{BL}}'_1(d)} (\mu(f) - \nu(f)) &\leq \sup_{f \in \underline{\mathrm{BL}}_1(d)} (\mu(f) - \nu(f)) \\ &\leq \int_\Omega \sup_{f \in \underline{\mathrm{BL}}_1(d)} (\mu_\omega(f(\omega, .)) - \nu_\omega(f(\omega, .))) \, d\mathrm{P}(\omega) \end{aligned}$$

$$\leq \int_\Omega D_{\mathrm{BL}(d)}(\mu_\omega, \nu_\omega)\, d\,\mathrm{P}(\omega).$$

Let us prove the converse inequalities. Let $\epsilon \in]0,1]$. The measures $\mu(\Omega\times .)$ and $\nu(\Omega\times .)$ are inner regular w.r.t. the totally bounded subsets of $\mathbb{S}$, thus there exists a totally bounded subset K of $\mathbb{S}$ such that

$$\text{(2.2.2)} \qquad \mu(\Omega\times K) \geq 1-\epsilon \quad \text{and} \quad \nu(\Omega\times K) \geq 1-\epsilon.$$

As in the proof of Lemma 2.1.1, for any continuous function f on $\mathbb{S}$ and any $B \subset \mathbb{S}$, let us denote $\|f\|_B := \sup_{x\in B} |f(x)|$. The set of restrictions to K of elements of $\mathrm{BL}_1(d)$ is totally bounded for $\|.\|_K$, thus there exist $h_1,\dots,h_n \in \mathrm{BL}_1(d)$ such that, for each $h \in \mathrm{BL}_1(d)$, we have $\inf_{i=1,\dots,n} \|h-h_i\|_K \leq \epsilon$. For every $\omega\in\Omega$, there exists $N(\omega) \in \{1,\dots,n\}$ such that

$$\mu_\omega(h_{N(\omega)}\,\mathbf{1}_K) - \nu_\omega(h_{N(\omega)}\,\mathbf{1}_K) \geq \sup_{h\in\mathrm{BL}_1(d)} \left(\mu_\omega(h\,\mathbf{1}_K) - \nu_\omega(h\,\mathbf{1}_K)\right) - 2\epsilon.$$

Obviously, we can assume that N is measurable. We have, for every $\omega\in\Omega$,

$$\begin{aligned} D_{\mathrm{BL}(d)}(\mu_\omega,\nu_\omega) &\leq \mu_\omega(\mathbb{S}\setminus K) \\ &\quad + \nu_\omega(\mathbb{S}\setminus K) + \sup_{h\in\mathrm{BL}_1(d)} \left(\mu_\omega(h\,\mathbf{1}_K) - \nu_\omega(h\,\mathbf{1}_K)\right) \\ \text{(2.2.3)} \qquad &\leq \mu_\omega(\mathbb{S}\setminus K) + \nu_\omega(\mathbb{S}\setminus K) \\ &\quad + \mu_\omega(h_{N(\omega)}\,\mathbf{1}_K) - \nu_\omega(h_{N(\omega)}\,\mathbf{1}_K) + 2\epsilon \\ &\leq 2\mu_\omega(\mathbb{S}\setminus K) \\ &\quad + 2\nu_\omega(\mathbb{S}\setminus K) + \mu_\omega(h_{N(\omega)}) - \nu_\omega(h_{N(\omega)}) + 2\epsilon. \end{aligned}$$

Using (2.2.2) and (2.2.3), we thus have

$$\begin{aligned} \int_\Omega D_{\mathrm{BL}(d)}(\mu_\omega,\nu_\omega)\, d\,\mathrm{P}(\omega) &\leq 2\mu(\Omega\times K^c) + 2\nu(\Omega\times K^c) \\ &\quad + \int_\Omega \mu_\omega(h_{N(\omega)}) - \nu_\omega(h_{N(\omega)})\, d\,\mathrm{P}(\omega) + 2\epsilon \\ &\leq \sup_{f\in\underline{\mathrm{BL}}'_1(d)} (\mu(f) - \nu(f)) + 6\epsilon \end{aligned}$$

because the mapping $\Omega\times\mathbb{S} \mapsto h_{N(\omega)}(x)$ is in $\underline{\mathrm{BL}}'_1(d)$. As ϵ is arbitrary, this shows that we have

$$\int_\Omega D_{\mathrm{BL}(d)}(\mu_\omega, \nu_\omega)\, d\mathrm{P}(\omega) = \sup_{f \in \underline{\mathrm{BL}}'_1(d)} (\mu(f) - \nu(f)).$$

□

Remark. Let $(\mathbb{S}, d)$ be a metric space. From Lemma 2.1.1 and Theorem 2.2.1, stable convergence in $\mathcal{Y}(\Omega, \mathcal{S}, \mathrm{P}; \mathbb{S})$ amounts to pointwise convergence on $\underline{\mathrm{BL}}_1(d)$ (or on $\underline{\mathrm{BL}}(d)$), whereas convergence in probability amounts to uniform convergence on $\underline{\mathrm{BL}}_1(d)$. Thus, in $\mathcal{Y}(\Omega, \mathcal{S}, \mathrm{P}; \mathbb{S})$, convergence in probability implies stable convergence.

Corollary 2.2.2. *Let $(\mathbb{S}, d)$ be a metric space. Let $(\mu^\alpha)_{\alpha \in \mathbb{A}}$ be a net in $\mathcal{Y}(\Omega, \mathcal{S}, \mathrm{P}; \mathbb{S})$ and let $X \in L^0(\Omega; \mathbb{S})$. Then (μ^α) converges in probability to $\underline{\delta}_X$ if and only if (μ^α) stably converges to $\underline{\delta}_X$.*

Proof. We already know that convergence in probability implies stable convergence.

Assume that $(\mu^\alpha)_{\alpha \in \mathbb{A}}$ stably converges to $\mu = \underline{\delta}_X$. Let $f \in \underline{\mathrm{BL}}_1(d)$. We have

$$\begin{aligned}
&\left| \int f(\omega, x)\, d(\mu^\alpha_\omega - \mu_\omega)(x)\, d\mathrm{P}(\omega) \right| \\
&\quad = \left| \int \left(\int f(\omega, x) - f(x, X(\omega))\, d\mu^\alpha_\omega(x) \right) d\mathrm{P}(\omega) \right| \\
&\quad \le \int \left(\int |f(\omega, x) - f(\omega, X(\omega))|\, d\mu^\alpha_\omega(x) \right) d\mathrm{P}(\omega) \\
&\quad \le \int d(x, X(\omega)) \wedge 1\, d\mu^\alpha(\omega, x).
\end{aligned}$$

Let $g \in \underline{\mathrm{BL}}(d)$ be defined by $g(\omega, x) = d(x, X(\omega)) \wedge 1$. As $(\mu^\alpha)_{\alpha \in \mathbb{A}}$ stably converges to $\mu = \underline{\delta}_X$, we have, using Lemma 2.1.1,

$$\lim_\alpha \int g\, d\mu^\alpha = \int g\, d\mu = \int d(X, X)\, d\mathrm{P} = 0,$$

and thus

$$\begin{aligned}
&\sup_{f \in \underline{\mathrm{BL}}_1(d)} \int f(\omega, x)\, d(\mu^\alpha_\omega - \mu_\omega)(x)\, d\mathrm{P}(\omega) \\
&\quad \le \left| \int g(\omega, x)\, d(\mu^\alpha_\omega - \mu_\omega)(x)\, d\mathrm{P}(\omega) \right| \to 0.
\end{aligned}$$

□

Proposition 2.2.3. *Assume that $\mathbb{S}$ is Polish and that $\mathcal{S}$ is countably generated. Let $\mathcal{H}$ be a convex subset of $\mathcal{Y}(\Omega, \mathcal{S}, \mathrm{P}; \mathbb{S})$. Then $\mathcal{H}$ is closed for the stable topology if and only if it is closed for the topology of convergence in probability.*

Proof. If $\mathcal{H}$ is closed for the stable topology, it is obviously closed for the topology of convergence in probability, which is finer. Assume now that $\mathcal{H}$ is closed for the topology of convergence in probability. It is well known that, as $\mathcal{S}$ is countably generated, the stable topology is metrizable (see e.g. [Bal00, Theorem 4.6 page 35]). We thus only need to prove that $\mathcal{H}$ is sequentially closed. Let $(\mu^n)_n$ be a sequence in $\mathcal{H}$ which stably converges to some $\mu \in \mathcal{Y}(\Omega, \mathcal{S}, \mathrm{P}; \mathbb{S})$. From Balder's result on K–convergence [Bal89 Bal90] (or we could instead use the Mazur property), there exists a subsequence $(\nu_n)_n$ of $(\mu^n)_n$ such that

$$\lim_n \frac{1}{n} \sum_{i=1}^{n} \nu_\omega^i = \mu_\omega \text{ a.e.}$$

Thus the sequence $(1/n \sum_{i=1}^n \nu^i)_n$ converges in probability to μ. As $\mathcal{H}$ is convex and closed in probability, we thus have $\mu \in \mathcal{H}$. □

2.3 Fiber product lemma

Let $\mathbb{S}$ and $\mathbb{T}$ be metric spaces and let $\mu \in \mathcal{Y}(\Omega, \mathcal{S}, \mathrm{P}; \mathbb{S})$ and $\nu \in \mathcal{Y}(\Omega, \mathcal{S}, \mathrm{P}; \mathbb{T})$. We call *fiber product* of μ and ν the Young measure $\mu \underline{\otimes} \nu \in \mathcal{Y}(\Omega, \mathcal{S}, \mathrm{P}; \mathbb{S} \times \mathbb{T})$ defined by

$$(\mu \underline{\otimes} \nu)_\omega = \mu_\omega \otimes \nu_\omega$$

for every $\omega \in \Omega$.

The following theorem generalizes a classical result [Fis70 Bal88 Val90 Val94 Tat02]. In these papers (except [Tat02]), $(\mu^\alpha)_\alpha$ stably converges to a degenerate Young measure, but, from Corollary 2.2.2, this assumption implies Hypothesis (i) below.

Theorem 2.3.1. (Fiber product lemma) *Let $(\mathbb{S}, d_\mathbb{S})$ and $(\mathbb{T}, d_\mathbb{T})$ be metric spaces. Let $(\mu^\alpha)_{\alpha \in \mathbb{A}}$ be a net in $\mathcal{Y}(\Omega, \mathcal{S}, \mathrm{P}; \mathbb{S})$ and $(\nu^\alpha)_{\alpha \in \mathbb{A}}$ be a net in $\mathcal{Y}(\Omega, \mathcal{S}, \mathrm{P}; \mathbb{T})$ (with same index set). Assume that*

(i) $(\mu^\alpha)_\alpha$ converges in probability to $\mu^\infty \in \mathcal{Y}(\Omega, \mathcal{S}, \mathrm{P}; \mathbb{S})$,
(ii) $(\nu^\alpha)_\alpha$ stably converges to $\nu^\infty \in \mathcal{Y}(\Omega, \mathcal{S}, \mathrm{P}; \mathbb{T})$.
Then $(\mu^\alpha \underline{\otimes} \nu^\alpha)_\alpha$ stably converges to $\mu^\infty \underline{\otimes} \nu^\infty$.

First proof. For all $(s,t), (s',t') \in \mathbb{S} \times \mathbb{T}$, set

$$d((s,t),(s',t')) = \max \{ d_\mathbb{S}(s,s'), d_\mathbb{T}(t,t') \}.$$

Let $A \in \mathcal{S}$ and let $f : \mathbb{S} \times \mathbb{T} \to [0,1]$ be an element of $\mathrm{BL}_1(\mathbb{S} \times \mathbb{T}, d)$. For each $\alpha \in \mathbb{A} \cup \{\infty\}$, each $\omega \in \Omega$ and each $s \in \mathbb{S}$, let

$$g^\alpha(\omega, s) = \mathbf{1}_A(\omega) \int_\mathbb{T} f(s,t)\, d\nu_\omega^\alpha(t).$$

Then each g^α is in $\underline{\mathrm{BL}}_1(d_{\mathbb{S}})$, thus, from (i) and Theorem 2.2.1,

$$\limsup_{\alpha} \sup_{\beta\in\mathbb{A}} \left| \int_{\Omega\times\mathbb{S}} g^\beta \, d(\mu^\alpha - \mu^\infty) \right| = 0.$$

In particular, we have

$$(2.3.1) \qquad \begin{aligned} &\lim_\alpha \int \mathbf{1}_A(\omega) f(s,t)\, d(\mu^\alpha_\omega - \mu^\infty_\omega)(s)\, d\nu^\alpha_\omega(t)\, d\mathrm{P}(\omega) \\ &= \lim_\alpha \int_{\Omega\times\mathbb{S}} g^\alpha \, d(\mu^\alpha - \mu^\infty) \\ &= 0. \end{aligned}$$

Set $h(\omega,t) = \mathbf{1}_A(\omega) \int f(s,t)\, d\mu^\infty_\omega(s)$ for all $(\omega,t) \in \Omega \times \mathbb{T}$. Then h is in $\underline{\mathrm{BL}}_1(d_{\mathbb{T}})$, thus, from (ii) and Lemma 2.1.1, we also have

(2.3.2)

$$\lim_\alpha \int \mathbf{1}_A(\omega) f(s,t)\, d\mu^\infty_\omega(s)\, d\nu^\alpha_\omega(t)\, d\mathrm{P}(\omega) = \lim_\alpha \int h\, d\nu^\alpha = \int h\, d\nu^\infty$$
$$= \int \mathbf{1}_A(\omega) f(s,t)\, d\left(\mu^\infty \underline{\otimes} \nu^\infty\right)(\omega,s,t).$$

Using (2.3.1) and (2.3.2), we immediately get

$$\begin{aligned} &\lim_\alpha \int \mathbf{1}_A(\omega) f(s,t)\, d\left(\mu^\alpha \underline{\otimes} \nu^\alpha\right)(\omega,s,t) \\ &\qquad = \lim_\alpha \int \mathbf{1}_A(\omega) f(s,t)\, d(\mu^\alpha_\omega - \mu^\infty_\omega)(s)\, d\nu^\alpha_\omega(t)\, d\mathrm{P}(\omega) \\ &\qquad\quad + \lim_\alpha \int \mathbf{1}_A(\omega) f(s,t)\, d\mu^\infty_\omega(s)\, d\nu^\alpha_\omega(t)\, d\mathrm{P}(\omega) \\ &\qquad = \int \mathbf{1}_A(\omega) f(s,t)\, d(\mu^\infty \underline{\otimes} \nu^\infty)(\omega,s,t), \end{aligned}$$

which proves that $(\mu^\alpha \underline{\otimes} \nu^\alpha)_{\alpha\in\mathbb{A}}$ stably converges to $\mu^\infty \underline{\otimes} \nu^\infty$. □

Second proof of Theorem 2.3.1.. Let us define the distance d on $\mathbb{S}\times\mathbb{T}$ as in the first proof. From Corollary 2.1.2, Condition (i) is equivalent to the convergence in measure in $L^\infty_{\mathrm{BL}(\mathbb{S})^*}(\Omega, \mathcal{S}, \mathrm{P})$. Let $f \in L^1_{\mathrm{BL}(\mathbb{S}\times\mathbb{T},d)}$. The mapping $r : \omega \to \|f(\omega,.)\|_{\mathrm{BL}(d)}$ is in $L^1_{\mathbb{R}}(\Omega, \mathcal{S}, \mathrm{P})$. It is easy to see that the mappings $\omega \mapsto g^\alpha_\omega$ where

$$g^\alpha_\omega(s) = \int_{\mathbb{T}} f_\omega(s,t)\, d\nu^\alpha_\omega(t) = \langle f_\omega(s,.), \nu^\alpha_\omega \rangle_{\langle \mathcal{C}(\mathbb{T}), \mathcal{C}(\mathbb{T})^* \rangle}$$

belong to $L^1_{\mathrm{BL}(\mathbb{S})}$; further it is obvious that the net (g^α) is uniformly integrable in $L^1_{\mathrm{BL}(\mathbb{S})}$ because $\|g^\alpha_\omega\|_{\mathrm{BL}(d_\mathbb{S})} \leq r(\omega)$ for all $\omega \in \Omega$. By (i) and the above embedding theorem, the net $(\mu^\alpha - \mu^\infty)$ converges to 0 in measure in $\mathrm{BL}(\mathbb{S})^*$, as this net is uniformly bounded in $L^\infty_{\mathrm{BL}(\mathbb{S})^*}$. From [Cas80], we see that $(\mu^\alpha - \mu^\infty)$ converges to 0 uniformly on the uniformly integrable set $\mathcal{H} := \{g^\alpha\} \subset L^1_{\mathrm{BL}(\mathbb{S})}(\Omega, \mathcal{S}, \mathrm{P})$. In particular, we have,

$$\lim_{\alpha \to \infty} \int_\Omega \langle g^\alpha_\omega, \mu^\alpha_\omega - \mu^\infty_\omega \rangle \, d\,\mathrm{P}(\omega) = 0, \tag{2.3.3}$$

here the bracket $\langle .,. \rangle$ denotes the duality $\langle \mathrm{BL}(\mathbb{S}), \mathrm{BL}(\mathbb{S})^* \rangle$. Set

$$h^\infty(\omega, t) := \int_\mathbb{S} f_\omega(s,t) d\mu^\infty_\omega(s)$$

for all $(\omega, t) \in \Omega \times \mathbb{T}$. Then by (ii) we immediately have

$$\lim_{n \to \infty} \int_\Omega \langle h^\infty_\omega, \nu^\alpha_\omega \rangle \, d\,\mathrm{P}(\omega) = \int_\Omega \langle h^\infty_\omega, \nu^\infty_\omega \rangle \, d\,\mathrm{P}(\omega) \tag{2.3.4}$$

From (2.3.3) and (2.3.4) the result follows. □

Remarks. 1) With the notations used in both proofs of Theorem 2.3.1, it is not difficult to prove that, if $\mu, \mu' \in \mathcal{Y}(\Omega, \mathcal{S}, \mathrm{P}; \mathbb{S})$ and $\nu, \nu' \in \mathcal{Y}(\Omega, \mathcal{S}, \mathrm{P}; \mathbb{T})$, then

$$\Delta_{\underline{\mathrm{BL}}(d)}(\mu \underline{\otimes} \nu, \mu' \underline{\otimes} \nu') \leq \Delta_{\underline{\mathrm{BL}}(d_\mathbb{S})}(\mu, \mu') + \Delta_{\underline{\mathrm{BL}}(d_\mathbb{T})}(\nu, \nu').$$

Thus, with the hypothesis of Theorem 2.3.1, if furthermore $(\nu^\alpha)_\alpha$ converges in probability, then $(\mu^\alpha \underline{\otimes} \nu^\alpha)_\alpha$ converges in probability to $\mu^\infty \underline{\otimes} \nu^\infty$.

2) The preceding proposition holds in the particular case when we assume in (i) that (ν^α) is a *sequence* (ν^n) which converges pointwisely on Ω to $\nu^\infty \in \mathcal{Y}(\Omega, \mathcal{S}, \mathrm{P}; \mathbb{S})$. At this point, an alternative proof (see e.g. [Tat02]) (for sequences instead of nets, without using the Dudley embedding theorem and [Cas80]) is available, by considering the "augmented" Young measures $(\delta_n \otimes \nu^n)$ in $\mathcal{Y}(\Omega, \mathcal{S}, \mathrm{P}; \hat{\mathbb{N}} \times \mathbb{S})$.

For the sake of completeness, let us mention a Vitali-Young convergence type theorem which is an easy consequence of Theorem 2.3.1 that we need in the statement of next results.

Corollary 2.3.2. *Let $\mathbb{S}$ and $\mathbb{T}$ be metric spaces. Let $(u^\alpha)_{\alpha \in \mathbb{A}}$ be a net of $\mathcal{S}$-measurable mappings from Ω into $\mathbb{S}$ which converges in probability to a $\mathcal{S}$-measurable mapping u^∞ from Ω into $\mathbb{S}$ and let $(v^\alpha)_{\alpha \in \mathbb{A}}$ be a net of $\mathcal{S}$-measurable mappings from Ω into $\mathbb{T}$ such that (v^α) stably converges to a Young* *measure* $\nu^\infty \in \mathcal{Y}(\Omega, \mathcal{S}, \mathrm{P}; \mathbb{T})$. *Let* $h : \Omega \times \mathbb{S} \times \mathbb{T} \to \mathbb{R}$ *be a Carathéodory integrand such that the net* $(h(., u^\alpha(.), v^\alpha(.))$ *is uniformly integrable. Then, the following holds*

$$\lim_\alpha \int_\Omega h(\omega, u^\alpha(\omega), v^\alpha(\omega))\, d\mathrm{P}(\omega) = \int_\Omega [\int_\mathbb{T} h(\omega, u^\infty(\omega), t)\, d\nu_\omega^\infty(t)]\, d\mathrm{P}(\omega).$$

Proof. If h is bounded, the result is immediate, because, from Theorem 2.3.1, $(\underline{\delta}_{u^\alpha} \otimes \underline{\delta}_{v^\alpha})$ stably converges to $\underline{\delta}_{u^\infty} \otimes \nu^\infty$.

Now, for the general case, we have $h = h^+ - h^-$ and the nets $(h^+(., u^\alpha(.), v^\alpha(.))$ and $(h^-(., u^\alpha(.), v^\alpha(.))$ are uniformly integrable. We thus only need to prove the result for h^+ and h^-, that is, we can assume w.l.g. that h is nonnegative.

Let $\epsilon > 0$. There exists $a_\epsilon > 0$ such that

$$\sup_{\alpha \in \mathbb{A}} \int_{\{|h(., u^\alpha(.), v^\alpha(.))| > a_\epsilon\}} |h(\omega, u^\alpha(\omega), v^\alpha(\omega))|\, d\mathrm{P}(\omega) \le \epsilon. \tag{2.3.5}$$

On the other hand, by Beppo-Levi Theorem, we can take a_ϵ large enough such that

$$\int_\Omega [\int_\mathbb{T} \mathbf{1}_{\{h(\omega, u^\infty(\omega), .) > a_\epsilon\}}(t)\, h(\omega, u^\infty(\omega), t)\, d\nu_\omega^\infty(t)]\, d\mathrm{P}(\omega) \le \epsilon. \tag{2.3.6}$$

Let us define a continuous truncation function $\alpha_\epsilon : \mathbb{R}^+ \to \mathbb{R}^+$ such that

$$\begin{cases} \alpha_\epsilon(x) = 0 & \text{if } x \ge a_\epsilon + 1, \\ 0 \le \alpha_\epsilon(x) \le a_\epsilon & \text{if } a_\epsilon \le x \le a_\epsilon + 1, \\ \alpha_\epsilon(x) = x & \text{if } 0 \le x \le a_\epsilon. \end{cases}$$

Let $h_\epsilon = \alpha_\epsilon \circ h$. Then h_ϵ is a bounded Carathéodory integrand, thus we have

$$\lim_\alpha \int_\Omega h_\epsilon(\omega, u^\alpha(\omega), v^\alpha(\omega))\, d\mathrm{P}(\omega) = \int_\Omega [\int_\mathbb{T} h_\epsilon(\omega, u^\infty(\omega), t)\, d\nu_\omega^\infty(t)]\, d\mathrm{P}(\omega).$$

There exists thus $\alpha_0 \in \mathbb{A}$ such that, for all $\alpha \ge \alpha_0$,

(2.3.7)

$$\left| \int_\Omega h_\epsilon(\omega, u^\alpha(\omega), v^\alpha(\omega))\, d\mathrm{P}(\omega) - \int_\Omega [\int_\mathbb{T} h_\epsilon(\omega, u^\infty(\omega), t)\, d\nu_\omega^\infty(t)]\, d\mathrm{P}(\omega) \right| \le \epsilon.$$

But, from (2.3.5) and (2.3.6), we also have, for every $\alpha \ge \alpha_0$,

$$(2.3.8) \qquad \int_\Omega h(\omega, u^\alpha(\omega), v^\alpha(\omega))\, d\,\mathrm{P}(\omega) - \int_\Omega h_\epsilon(\omega, u^\alpha(\omega), v^\alpha(\omega))\, d\,\mathrm{P}(\omega) \leq \epsilon,$$

$$\int_\Omega [\int_\mathbb{T} h(\omega, u^\infty(\omega), t)\, d\nu_\omega^\infty(t)]\, d\,\mathrm{P}(\omega) - \int_\Omega [\int_\mathbb{T} h_\epsilon(\omega, u^\infty(\omega), t)\, d\nu_\omega^\infty(t)]\, d\,\mathrm{P}(\omega) \leq \epsilon.$$

Adding (2.3.8) to (2.3.7) yields, for $\alpha \geq \alpha_0$,

$$\left|\int_\Omega h(\omega, u^\alpha(\omega), v^\alpha(\omega))\, d\,\mathrm{P}(\omega) - \int_\Omega [\int_\mathbb{T} h(\omega, u^\infty(\omega), t)\, d\nu_\omega^\infty(t)]\, d\,\mathrm{P}(\omega)\right| \leq 3\epsilon.$$

□

3. Some new applications of the fiber product lemma for Young measures

In the sequel, $\mathbb{E} = \mathbb{R}^d$ is a finite dimensional space and $[0, 1]$ is equipped with the Lebesgue measure. The closed unit ball of $\mathbb{E}$ is denoted by $\bar{B}_E(0, 1)$.

3.1 The value function of a control problem governed by a first order ordinary differential equation

In this section we present a study of the value function of a control problem where the controls are Young measures. As the proofs are rather long, we do not make weak assumptions on the Control spaces but we only focus on the main ideas in order to present some sharp applications of the fiber product for Young measures presented above. Namely we assume here that $\mathbb{S}$ and $\mathbb{Z}$ are metric compact spaces. Let $k(\mathbb{Z})$ be the set of all compact subsets of $\mathbb{Z}$, $\Gamma : [0, 1] \to k(\mathbb{Z})$ be a compact valued Lebesgue measurable multifunction from $[0, 1]$ to $\mathbb{Z}$. Let $\mathcal{M}_+^1(\mathbb{S})$ (resp. $\mathcal{M}_+^1(\mathbb{Z})$) be the set of all probability Radon measures on $\mathbb{S}$ (resp. $\mathbb{Z}$). It is well known that $\mathcal{M}_+^1(\mathbb{S})$ (resp. $\mathcal{M}_+^1(\mathbb{Z})$) is a compact metrizable space for the $\sigma(\mathcal{C}(\mathbb{S})^*, \mathcal{C}(\mathbb{S}))$ (resp. $\sigma(\mathcal{C}(\mathbb{Z})^*, \mathcal{C}(\mathbb{Z}))$–topology.

Let us consider a mapping $f : [0, 1] \times \mathbb{E} \times \mathbb{S} \times \mathbb{Z} \to \mathbb{E}$ satisfying:

(i) for every fixed $t \in [0, 1]$, $f(t, ., ., .)$ is continuous on $\mathbb{E} \times \mathbb{S} \times \mathbb{Z}$;

(ii) for every $(x, s, z) \in \mathbb{E} \times \mathbb{S} \times \mathbb{Z}$, $f(., x, s, z)$ is Lebesgue-measurable on $[0, 1]$;

(iii) there is a positive Lebesgue integrable function c such that $f(t, x, s, z) \in c(t)\bar{B}_E(0, 1)$ for all (t, x, s, z) in $[0, 1] \times \mathbb{E} \times \mathbb{S} \times \mathbb{Z}$;

(iv) there exists a Lipschitz constant λ such that

$$||f(t, x_1, s, z) - f(t, x_2, s, z)|| \leq \lambda ||x_1 - x_2||$$

for all $(t, x_1, s, z), (t, x_2, s, z) \in [0, 1] \times \mathbb{E} \times \mathbb{S} \times \mathbb{Z}$.

We consider the absolutely continuous solutions set of the following ordinary differential equations (ODE)

$$(\mathcal{I}_{M,\mathcal{H},\mathcal{O}}) \begin{cases} \dot{u}_{x,\mu,\zeta}(t) = \int_{\mathbb{S}} f(t, u_{x,\mu,\zeta}(t), s, \zeta(t))\, d\mu_t(s) \\ u_{x,\mu,\zeta}(0) = x \in M \subset \mathbb{E} \end{cases}$$

where ζ belongs to the set S_Γ of all original controls, which means that ζ is a Lebesgue-measurable mapping from $[0, 1]$ into $\mathbb{Z}$ with $\zeta(t) \in \Gamma(t)$ for a.e. $t \in [0, 1]$, and M is a compact subset of $\mathbb{E}$, $\mu \in \mathcal{H}$, where $\mathcal{H}$ is a subset in the space of Young measures $\mathcal{Y}([0, 1]; \mathbb{S})$ defined on $\mathbb{S}$, and

$$(\mathcal{I}_{M,\mathcal{H},\mathcal{R}}) \begin{cases} \dot{u}_{x,\mu,\nu}(t) = \int_{\Gamma(t)} [\int_{\mathbb{S}} f(t, u_{x,\mu,\nu}(t), s, z)\, d\mu_t(s)]\, d\nu_t(z) \\ u_{x,\mu,\nu}(0) = x \in M \subset \mathbb{E} \end{cases}$$

where ν belongs to the set $\mathcal{R}$ of all relaxed controls, which means that ν is a Lebesgue-measurable selection of the multifunction Σ defined by

$$\Sigma(t) := \{\sigma \in \mathcal{M}^1_+(\mathbb{Z});\ \sigma(\Gamma(t)) = 1\}$$

for all $t \in [0, 1]$. These assumptions are sufficient to guarantee that for each $(x, \mu, \nu) \in M \times \mathcal{H} \times \mathcal{R}$ there is a unique absolutely continuous solution $u_{x,\mu,\nu}$ for the ODE under consideration on the interval $[0, 1]$ with $u_{x,\mu,\nu}(0) = x \in M \subset \mathbb{E}$. Now comes a Bolza-type optimal control problem associated with the preceding ODE.

Theorem 3.1.1. *Assume that $J : [0, 1] \times \mathbb{E} \times \mathbb{S} \times \mathbb{Z} \to \mathbb{R}$ is an L^1-bounded Carathéodory integrand, (that is, $J(t, ., ., .)$ is continuous on $\mathbb{E} \times \mathbb{S} \times \mathbb{Z}$ for every $t \in [0, 1]$ and $J(., x, s, z)$ is Lebesgue-measurable on $[0, 1]$, for every $(x, s, z) \in \mathbb{E} \times \mathbb{S} \times \mathbb{Z}$) which satisfies the condition: there is an integrable function $\varphi \in L^1_{\mathbb{R}+}([0, 1]$ such that $|J(t, x, s, z)| \leq \varphi(t)$ for all $t, x, s, z) \in [0, 1] \times \mathbb{E} \times \mathbb{S} \times \mathbb{Z}$. Assume further that $\mathcal{H}$ is compact for the convergence in probability. Let us consider the control problems*

$$(P_{M,\mathcal{H},\mathcal{O}}) : \inf_{(x,\mu,\zeta)\in M\times\mathcal{H}\times S_\Gamma} \int_0^1 [\int_{\mathbb{S}} J(t, u_{x,\mu,\zeta}(t), s, \zeta(t))\, d\mu_t(s)]\, dt$$

and

$$(P_{M,\mathcal{H},\mathcal{R}}) : \inf_{(x,\mu,\nu)\in M\times\mathcal{H}\times\mathcal{R}} \int_0^1 [\int_{\mathbb{Z}} [\int_{\mathbb{S}} J(t, u_{x,\mu,\nu}(t), s, z)\, d\mu_t(s)]\, d\nu_t(z)]\, dt$$

where $u_{x,\mu,\zeta}$ *(resp.* $u_{x,\mu,\nu}$*) is the unique solution associated with* (x, μ, ζ) *(resp.* (x, μ, ν) *to the ODE* $(\mathcal{I}_{M,\mathcal{O}})$ *(resp.* $(\mathcal{I}_{M,\mathcal{R}})$*. Then one has* $\inf(P_{M,\mathcal{H},\mathcal{O}}) = \min(P_{M,\mathcal{H},\mathcal{R}})$.

Proof. Claim 1. The graph of the single valued mapping $(x, \mu, \nu) \mapsto u_{x,\mu,\nu}$ defined on the compact space $M \times \mathcal{H} \times \mathcal{R}$ with value in the Banach space $C_{\mathbb{E}}([0,1])$ endowed with the topology of the sup-norm is compact. It is obvious that the solution $u_{x,\mu,\nu}$ for the ODE under consideration is given explicitely by

$$u_{x,\mu,\nu}(t) = x + \int_0^t [\int_{\mathbb{Z}} [\int_S f(\tau, u_{x,\mu,\nu}(\tau), s, z)\, d\mu_\tau(s)]\, d\nu_\tau(z)]dt$$

for each $t \in [0,1]$. Let (x^n, μ^n, ν^n) be a sequence in $M \times \mathcal{H} \times \mathcal{R}$ and let u_{x^n,μ^n,ν_n} be the unique absolutely continuous solution to

$$\begin{cases} \dot{u}_{x^n,\mu^n,\nu^n}(t) = \int_{\mathbb{Z}} [\int_{\mathbb{S}} f(t, u_{x^n,\mu^n,\nu^n}(t), s, z)\, d\mu^n_t(s)]\, d\nu^n_t(z) \ a.e. \\ u_{x^n,\mu^n,\nu^n}(0) = x^n \in M. \end{cases}$$

Since M is compact we may suppose that (x^n) converges to a point $x^\infty \in M$. Taking into account the assumption on f, it is easily seen that the sequence (u_{x^n,μ^n,ν^n}) is relatively compact in $C_{\mathbb{E}}([0,1])$. We may suppose, by extracting subsequences, that (u_{x^n,μ^n,ν^n}) converges uniformly to an absolutely continuous function $u^\infty(.)$ with $u^\infty(0) = x^\infty$ and $(\dot{u}_{x^n,\mu^n,\nu^n})$ converges $\sigma(L^1_{\mathbb{E}}, L^\infty_{\mathbb{E}})$ to $\dot{u}^\infty$. We may also assume that (μ^n) converges in probability to $\mu^\infty \in \mathcal{H}$. Further the sequence (ν^n) of Young measures is relatively compact for the stable topology on the space $\mathcal{Y}([0,1]; \mathbb{Z})$ of Young measures, and hence by extracting a subsequence, we may suppose that (ν^n) stably converges to a Young measure ν^∞ with $\nu^\infty_t(\Gamma(t)) = 1$ a.e. So, in view of Theorem 2.3.1, $\underline{\delta}_{u_{x^n,\mu^n,\nu^n}} \underline{\otimes}\, \mu^n \underline{\otimes}\, \nu^n$ stably converges to $\underline{\delta}_{u^\infty} \underline{\otimes}\, \mu^\infty \underline{\otimes}\, \nu^\infty$. Let $h \in L^\infty_{\mathbb{E}}([0,1])$. It is clear that the function $L : (t, x, s, z) \mapsto \langle h(t), f(t, x, s, z)\rangle$ is an L^1- bounded Carathéodory integrand defined on the compact space $[0,1] \times M \times \mathbb{S} \times \mathbb{Z}$, namely, $|L(t, x, s, z)| \leq h(t)c(t)$ for all $(t, x, s, z) \in [0,1] \times M \times \mathbb{S} \times \mathbb{Z}$, using condition (iii). Consequently, by the stable convergence of $(\underline{\delta}_{u_{x^n,\mu^n,\nu^n}} \underline{\otimes}\, \mu^n \underline{\otimes}\, \nu^n)$ to $\underline{\delta}_{u^\infty} \underline{\otimes}\, \mu^\infty \underline{\otimes}\, \nu^\infty$, we get

$$\begin{aligned}
&\lim_{n\to\infty}\int_0^1[\int_{\mathbb{Z}}[\int_{\mathbb{S}}\langle h(t), f(t,u_{x^n,\mu^n,\nu^n}(t),s,z)\rangle\,\mu_t^n(s)]\,d\nu_t^n(z)]\,dt\\
=&\lim_{n\to\infty}\int_0^1\langle h(t), v_n(t)\rangle\,dt\\
=&\int_0^1[\int_{\mathbb{Z}}[\int_{\mathbb{S}}\langle h(t), f(t,u^\infty(t),s,z)\rangle\,d\mu_t^\infty(s)]\,d\nu_t^\infty(z)]\,dt\\
=&\int_0^1\langle h(t), v_\infty(t)\rangle\,dt
\end{aligned}$$

where, for notational convenience,

$$v^n(t)=\int_{\mathbb{Z}}[\int_{\mathbb{S}} f(t,u_{x^n,\mu^n,\nu^n}(t),s,z)\,d\mu_t^n(s)]\,d\nu_t^n(z),\ \forall t\in[0,1],$$

and

$$v^\infty(t)=\int_{\mathbb{Z}}[\int_{\mathbb{S}} f(t,u^\infty(t),s,z)\,d\mu_t^\infty(s)]\,d\nu_t^\infty(z),\ \forall t\in[0,1].$$

Hence (v^n) weakly converges in $L^1_{\mathbb{E}}([0,1])$ to v^∞. Using the weak convergence in $L^1_{\mathbb{E}}([0,1])$ of $(\dot u_{x^n,\mu^n,\nu^n})$ to $\dot u^\infty$, and the preceding limit, we get

$$\dot u^\infty(t)=\int_{\mathbb{Z}}[\int_{\mathbb{S}} f(t,u^\infty(t),s,z)\,d\mu_t^\infty(s)]\,d\nu_t^\infty(z)\ a.e..$$

So, we have necessarily $u^\infty(.)=u_{x^\infty,\mu^\infty,\nu^\infty}(.)$, where $u_{x^\infty,\mu^\infty,\nu^\infty}$ is the unique absolutely continuous solution of the ODE $(\mathcal{I}_{M,\mathcal{H},\mathcal{R}})$ associated with $(x^\infty,\mu^\infty,\nu^\infty)$. Hence Claim 1 is proved.

Claim 2. $\inf(P_{M,\mathcal{H},\mathcal{O}})=\min(P_{M,\mathcal{H},\mathcal{R}})$.

As a consequence of Claim 1, the solutions set $\{u_{x,\mu,\nu}:(x,\mu,\nu)\in M\times\mathcal{H}\times\mathcal{R}\}$ is compact for the topology of uniform convergence. Since $\mathcal{O}$ is dense in $\mathcal{R}$ for the stable topology, it suffices to prove that the mapping

$$\Psi:(x,\mu,\nu)\to\int_0^1[\int_{\mathbb{Z}}[\int_S J(t,u_{x,\mu,\nu}(t),s,z)\,d\mu_t(s)]\,d\nu_t(z)]\,dt$$

is continuous on $M\times\mathcal{H}\times\mathcal{R}$. Let (x^n,μ^n,ν^n) be a sequence in $M\times\mathcal{H},\times\mathcal{R}$ such that (x^n) converges to $x\in M$ and (ν^n) converges in probability to $\mu\in\mathcal{H}$ and (ν^n) stably converges to $\nu\in\mathcal{R}$. Applying the result in Claim 1, shows that (u_{x^n,μ^n,ν^n}) converges uniformly to $u_{x,\mu,\nu}$ that is a solution of our ODE

$$\begin{cases} \dot{u}_{x,\mu,\nu}(t) = \int_{\mathbb{Z}} [\int_{\mathbb{S}} f(t, u_{x,\mu,\nu}(t), s, z)\, d\mu_t(s)]\, d\nu_t(z) a.e. \\ u_{x,\mu,\nu}(0) = x \in M. \end{cases}$$

This implies that $(\underline{\delta_{u_{x^n,\mu^n,\nu^n}}} \otimes \underline{\mu^n} \otimes \underline{\nu^n})$ stably converges to $(\underline{\delta_{u_{x,\mu,\nu}}} \otimes \underline{\mu} \otimes \underline{\nu})$. As J is an L^1-bounded Carathéodory integrand, it follows that

$$\lim_{n\to\infty} \int_0^1 [\int_{\mathbb{Z}} [\int_{\mathbb{S}} J(t, u_{x^n,\mu^n,\nu^n}(t), s, z)\, d\mu_t^n(s)]\, d\nu_t^n(z)]\, dt$$
$$= \int_0^1 [\int_{\mathbb{Z}} [\int_{\mathbb{S}} J(t, u_{x,\mu,\nu}(t), s, z)\, d\mu_t(s)]\, d\nu_t(z)]\, dt.$$

The proof is therefore complete. □

Remarks. In the course of the proof of Theorem 3.1.1 we have proven a significant property, namely, the continuous dependence of the trajectories $u_{x,\mu,\nu}$ of the dynamic under consideration with respect to the data $(x, \mu, \nu) \in M \times \mathcal{H} \times \mathcal{R}$. At this point it is worth to mention that the continuity property stated in the Claim 1 of the proof of Theorem 3.1.1 holds if we replace $\mathcal{R}$ by any compact (metrizable) subset $\mathcal{K}$ for the stable convergence of the space of Young measures defined on $\mathbb{Z}$. To illustrate this fact and before going further we present below a min-max type result for Young measures.

Proposition 3.1.2. *Let $\mathbb{S}$ be a Polish space and $\mathcal{H}$ be a convex subset in $\mathcal{Y}([0,1];\mathbb{S})$ which is closed for the convergence in probability (or for the stable topology, see Proposition 2.2.3). Let $\mathbb{T}$ be a Polish space and let $\mathcal{K}$ be a compact (metrizable) subset, for the stable convergence, of the space of Young measures $\mathcal{Y}([0,1];\mathbb{T})$. Let us consider a real-valued function $\Phi : \mathcal{H} \times \mathcal{K}$ such that, for every fixed $\mu \in \mathcal{H}$, $\Phi(\mu, .)$ is upper-semicontinuous on $\mathcal{K}$ and for every fixed $\nu \in \mathcal{K}$, $\Phi(., \nu)$ is convex lower-semicontinuous on $\mathcal{H}$. Then there exist a pair $(\widetilde{\mu}, \widetilde{\nu}) \in \mathcal{H} \times \mathcal{K}$ such that*

$$\max_{\nu\in\mathcal{K}} \min_{\mu\in\mathcal{H}} \Phi(\mu,\nu) \le \Phi(\widetilde{\mu}, \widetilde{\nu}) \le \min_{\mu\in\mathcal{H}} \max_{\nu\in\mathcal{K}} \Phi(\mu,\nu).$$

Proof. Let us set

$$p(\mu) := \max_{\nu\in\mathcal{K}} \Phi(\mu,\nu), \ \forall \mu \in \mathcal{H},$$

$$q(\nu) = \inf_{\mu\in\mathcal{H}} \Phi(\mu,\nu), \ \forall \nu \in \mathcal{K}.$$

Then $p(.)$ is convex lower semicontinuous on $\mathcal{H}$ and $q(.)$ is upper semicontinuous on $\mathcal{K}$. As $\mathcal{H}$ has the Mazur property (see Proposition 2.1.3), it is not difficult to provide an element $\widetilde{\mu} \in \mathcal{H}$ such that

$$p(\widetilde{\mu}) = \min_{\mu \in \mathcal{H}} p(\mu).$$

As $q(.)$ is upper semicontinuous for the stable topology on the compact set $\mathcal{K}$ there exists $\widetilde{\nu} \in \mathcal{K}$ such that

$$q(\widetilde{\nu}) = \max_{\nu \in \mathcal{K}} q(\nu).$$

So we get

$$q(\widetilde{\nu}) \leq \Phi(\widetilde{\mu}, \widetilde{\nu}) \leq p(\widetilde{\mu}).$$

□

Proposition 3.1.3. *Assume that the hypotheses of Theorem 3.1.1 are satisfied. Let $l : \mathbb{E} \to \mathbb{E}$ be a continuous mapping. Let $x \in M$ and $\theta \in]0, 1]$ be fixed. Then the control problem*

$$\Psi_x(\mu, \nu) := \int_0^1 [\int_{\mathbb{Z}} [\int_{\mathbb{S}} J(t, u_{x,\mu,\nu}(t), s, z)\, d\mu_t(s)]\, d\nu_t(z)]\, dt + l(u_{x,\mu,\nu}(\theta))$$

subject to

$$\dot{u}_{x,\mu,\nu}(t) = \int_{\Gamma(t)} [\int_{\mathbb{S}} f(t, u_{x,\mu,\nu}(t), s, z)\, d\mu_t(s)]\, d\nu_t(z)$$
$$a.e.\ t \in [0, 1];\ \ u_{x,\mu,\nu}(0) = x \in M$$

admits at least a quasi-saddle point $(\widetilde{\mu}, \widetilde{\nu}) \in \mathcal{H} \times \mathcal{R}$, that is

$$\max_{\nu \in \mathcal{R}} \min_{\mu \in \mathcal{H}} \Psi_x(\mu, \nu) \leq \Psi_x(\widetilde{\mu}, \widetilde{\nu}) \leq \min_{\mu \in \mathcal{H}} \max_{\nu \in \mathcal{R}} \Psi_x(\mu, \nu).$$

Proof. Taking Theorem 3.1.1 and its remarks into account, we see that the function Ψ_x is continuous on $\mathcal{H} \times \mathcal{R}$. Repeating the arguments of Proposition 3.1.2 and using the compactness assumption on $\mathcal{H}$ gives the result. □

Remark. The conclusions of Theorem 3.1.1 and Proposition 3.1.3 may fail if one only assumes that $\mathcal{H}$ is compact for the stable topology (instead of the topology of convergence in probability), because the fiber product theorem is not valid: by compactness, the sequence $\underline{\delta}_{u_{x^n,\mu^n,\nu^n}} \underline{\otimes}\ \mu^n \underline{\otimes}\ \nu^n$ has limit points in $\mathcal{Y}([0, 1]; C_{\mathbb{E}}([0, 1]) \times \mathbb{S} \times \mathbb{Z})$, but they do not necessarily have the form $\underline{\delta}_{u^\infty} \underline{\otimes}\ \mu^\infty \underline{\otimes}\ \nu^\infty$ (see [Val94] for a counterexample).

In the preceding dynamical system, we focus essentially on the continuous dependence of the solutions on the data $(x, \mu, \nu) \in M \times \mathcal{H} \times \mathcal{R}$. In order to illustrate our techniques, we develop some new properties of the value function $V_g(x, \mu, t)$ associated with the dynamical system under consideration as follows:

$$V_g(x, \mu, t) = \inf_{\nu \in \mathcal{R}} g(u_{x,\mu,\nu}(t))$$

with the data $(x, \mu, \nu) \in M \times \mathcal{H} \times \mathcal{R}$ where g is a bounded lower semicontinuous function defined on $\mathbb{E}$ and $t \in [0, 1]$. Recall that $u_{x,\mu,\nu}(t)$ is the value of the solution $u_{x,\mu,\nu}$ at t of the dynamic $(\mathcal{I}_{M,\mathcal{H},\mathcal{R}})$. The following result provides a lower semicontinuity property for the integrand

$$J : (g, x, \mu, t) \mapsto V_g(x, \mu, t)$$

defined on $\mathcal{I}_{\mathbb{E}} \times M \times \mathcal{H} \times [0, 1]$ where $\mathcal{I}_{\mathbb{E}}$ denotes the set of all bounded lower semicontinuous functions defined on $\mathbb{E}$.

Proposition 3.1.4. *Assume the same hypothesis as in Theorem 3.1.1. Let $g \in \mathcal{I}_{\mathbb{E}}$, let $(x, \mu) \in M \times \mathcal{H}$ and let $t \in [0, 1]$. Then for any an increasing sequence (g_i) of bounded Lipschitz functions defined on $\mathbb{E}$ converging pointwisely to g, for any sequence (t^i) converging to t, for any a sequence (x^i) in M converging to $x \in M$ and for any sequence (μ^i) in $\mathcal{H}$ converging in measure to μ,*

$$\liminf_i V_{g^i}(x^i, \mu^i, t^i) \geq V_g(x, \mu, t).$$

Proof. Note that, for every $s \in [0, 1]$, the functions $(x, \mu, \nu) \to g_i(u_{x,\mu,\nu}(s))$ are continuous (thus uniformly continuous) on the compact space $M \times \mathcal{H} \times \mathcal{R}$ because the mapping $(x, \mu, \nu) \to (u_{x,\mu,\nu})$ is continuous on $M \times \mathcal{H} \times \mathcal{R}$ as we have already proved in Theorem 3.1.1. Therefore, for each i, the mapping $(s, x, \mu, \nu) \to g_i(u_{x,\mu,\nu}(s))$ is continuous. On the other hand, for each i, there is $\nu^i \in \mathcal{R}$ with

$$V_{g^i}(x^i, \mu^i, t^i) = g_i(u_{x^i,\mu^i,\nu^i}(t^i))$$

where u_{x^i,μ^i,ν^i} is the solution of the dynamic $(\mathcal{I}_{M,\mathcal{H},\mathcal{R}})$, that is

$$\dot{u}_{x^i,\mu^i,\nu^i}(\tau) = \int_{\Gamma(\tau)} [\int_{\mathbb{S}} f(\tau, u_{x^i,\mu^i,\nu^i}(\tau), s, z)\, d\mu^i_\tau(s)]\, d\nu^i_\tau(z)$$

$$a.e.\ \tau \in [0, 1],$$

$$u_{x^i,\mu^i,\nu^i}(0) = x^i.$$

Then, there is a subsequence of (μ^i, ν^i) with the same notation such that $(\mu^i \underline{\otimes} \nu^i)$ stably converges to $\mu \underline{\otimes} \nu$ and such that (u_{x^i,μ^i,ν^i}) converges uniformly to $u_{x,\mu,\nu}$ which is the solution of this dynamic, using Theorem 3.1.1, that is

$$\dot{u}_{x,\mu,\nu}(\tau) = \int_{\Gamma(\tau)} [\int_{\mathbb{S}} f(\tau, u_{x,\mu,\nu}(\tau), s, z)\, d\mu_\tau(s)]\, d\nu_\tau(z) \ a.e.\ \tau \in [0,1],$$

$$u_{x,\mu,\nu}(0) = x.$$

So, for $i \geq k$, we immediately have the estimate

$$\begin{aligned} V_{g_i}(x^i, \mu^i, t^i) &= g_i(u_{x^i,\mu^i,\nu^i}(t^i)) \\ &\geq g_k(u_{x^i,\mu^i,\nu^i}(t^i)) - g(u_{x,\mu,\nu}(t)) + V_g(x,\mu,t). \end{aligned}$$

Let $\varepsilon > 0$. There is $k > 0$ such that $0 \leq g(u_{x,\mu,\nu}(t)) - g_k(u_{x,\mu,\nu}(t)) \leq \varepsilon$. As (g_i) is increasing, we have

$$\begin{aligned} \liminf_{i\to\infty} g_i(u_{x^i,\mu^i,\nu^i}(t^i)) &\geq \liminf_{i\to\infty} g_k(u_{x^i,\mu^i,\nu^i}(t^i)) \\ &= g_k(u_{x,\mu,\nu}(t)) \geq g(u_{x,\mu,\nu}(t)) - \varepsilon. \end{aligned}$$

The preceding inequality implies that

$$\liminf_{i\to\infty} V_{g_i}(x^i, \mu^i, t^i) = \liminf_{i\to\infty} g_i(u_{x^i,\mu^i,\nu^i}(t^i)) \geq V_g(x,\mu,t) - \varepsilon.$$

□

In the preceding result the data g is of simple nature. Now we give an extension to Proposition 3.1.4 concerning a relaxation property for the relaxed value function, when the data is the integral functional associated with a positive, bounded normal integrand. Namely, let $h : [0,1] \times \mathbb{E} \times \mathbb{S} \times \mathbb{Z} \to \mathbb{R}^+$ be a bounded normal $\mathcal{L}([0,1])$–$\mathcal{B}(\mathbb{E} \times \mathbb{S} \times \mathbb{Z})$–measurable integrand, that is, h is measurable and $h(t,.,.,.,.)$ is lower-semicontinuous on $\mathbb{E} \times \mathbb{S} \times \mathbb{Z}$ for every fixed $t \in [0,1]$, and let I_h be the integral functional on $\mathcal{H} \times \mathcal{R}$ given by

$$I_h(x,\mu,\nu) = \int_0^1 [\int_{\mathbb{Z}} [\int_{\mathbb{S}} h(t, u_{x,\mu,\nu}(t), s, z)\, \mu_t(s)]\, d\nu_t(z)]\, dt,$$

where $u_{x,\mu,\nu}$ is the unique solution to

$$\dot{u}_{x,\nu,\mu}(t) = \int_{\Gamma(t)} [\int_{\mathbb{S}} f(t, u_{x,\mu,\nu}(t), s, z)\, d\mu_t(s)]\, d\nu_t(z) \ a.e.\ t \in [0,1],$$

$$u_{x,\mu,\nu}(0) = x.$$

Proposition 3.1.5. *Let h be as above, and let $(x,\mu) \in M \times \mathcal{H}$. Let (h^i) be an increasing sequence of bounded, Carathéodory integrands defined on $[0,1] \times \mathbb{E} \times \mathbb{S} \times \mathbb{Z}$ such that $h = \sup_i h^i$. Let (x^i) be a sequence in M which converges* *to $x \in M$, let (μ^i) be a sequence in $\mathcal{H}$ which converges in probability to μ. Let us consider the value function*

$$\begin{aligned} W_h(x,\mu) &= \inf_{\nu\in\mathcal{R}} I_h(x,\mu,\nu) \\ &= \inf_{\nu\in\mathcal{R}} \int_0^1 [\int_{\mathbb{Z}} [\int_{\mathbb{S}} h(t,u_{x,\mu,\nu}(t),s,z)\,d\mu_t(s)]\,d\nu_t(z)]\,dt, \end{aligned}$$

where $u_{x,\mu,\nu}$ is the solution of the dynamic $(\mathcal{I}_{M,\mathcal{H},\mathcal{R}})$. We then have

$$\liminf_i W_{h^i}(x^i,\mu^i) \geq W_h(x,\mu),$$

where, for all i,

$$\begin{aligned} W_{h^i}(x^i,\mu^i) &:= \inf_{\zeta\in S_\Gamma} I_{h^i}(x^i,\mu^i,\zeta) \\ &= \inf_{\zeta\in S_\Gamma} \int_0^1 [\int_S h^i(t,u_{x^i,\mu^i,\zeta}(t),s,\zeta(t))\,d\mu^i_t(s)]\,dt. \end{aligned}$$

Remark. It is well known that a sequence such as (h^i) always exists. It is enough to combine the Baire-Hausdorff approximations with the Projection Theorem in [CV77] to obtain such approximates (h^i).

Proof of Proposition 3.1.5. By Theorem 3.1.1, for each i, we have

$$\begin{aligned} \inf_{\nu\in S_\Sigma} I_{h^i}(x,\mu,\nu) &= \inf_{\nu\in S_\Sigma} \int_0^1 [\int_{\mathbb{Z}} [\int_S h^i(t,u_{x,\mu,\nu}(t),s,z)\,d\mu_t(s)]\,d\nu_t(z)]\,dt \\ &= \inf_{\zeta\in S_\Gamma} \int_0^1 [\int_S h^i(t,u_{x,\mu,\zeta}(t),s,\zeta(t))\,d\mu_t(s)]\,dt \\ &= W_{h^i}(x,\mu). \end{aligned}$$

We have already observed that the mappings $(x,\mu,\nu) \mapsto u_{x,\mu,\nu}$ and $(x,\mu,\nu) \mapsto I_{h^i}(x,\mu,\nu)$ are continuous on the compact space $M \times \mathcal{H} \times \mathcal{R}$. For each i, there is $\nu^i \in S_\Sigma$ with

$$W_{h^i}(x^i,\mu^i) = I_{h^i}(x^i,\mu^i,\nu^i),$$

where u_{x^i,μ^i,ν^i} is the trajectory solution of $(\mathcal{I}_{M,\mathcal{H},\mathcal{R}})$. Then, as in the proof of Proposition 3.1.4, there is a subsequence with the same notation such that $(\mu^i \underline{\otimes} \nu^i)$ stably converges to $\mu \underline{\otimes} \nu$ and (u_{x^i,μ^i,ν^i}) converges uniformly to $u_{x,\mu,\nu}$ which is a trajectory solution with the initial condition $u_{x,\mu,\nu}(0) = x$. So, for $i \geq k$, we immediately have the estimate

$$W_{h^i}(x^i,\mu^i) = I_{h^i}(x^i,\mu^i,\nu^i) \geq I_{h^k}(x^i,\mu^i,\nu^i) - I_h(x,\mu,\nu) + W_h(x,\mu).$$

For every fixed $(x,\mu,\nu) \in M \times \mathcal{H} \times \mathcal{R}$, we have, by using the monotone convergence theorem,

$$\begin{aligned}
\sup_i I_{h^i}(x,\mu,\nu) &= \sup_i \int_0^1 \int_{\mathbb{Z}} [\int_{\mathbb{S}} h^i(t, u_{x,\mu,\nu}(t), s, z)\, d\mu_t(s)]\, d\nu_t(z)]\, dt \\
&= \int_0^1 \int_{\mathbb{Z}} [\int_{\mathbb{S}} h(t, u_{x,\mu,\nu}(t), s, z)\, d\mu_t(s)]\, d\nu_t(z)]\, dt \\
&= I_h(x,\mu,\nu).
\end{aligned}$$

Let $\varepsilon > 0$. There is $k > 0$ such that $0 \le I_h(x,\mu,\nu) - I_{h^k}(x,\mu,\nu) \le \varepsilon$. As (I_{h^i}) is increasing, we have

$$\liminf_{i\to\infty} I_{h^i}(x^i,\mu^i,\nu^i) \ge \liminf_{i\to\infty} I_{h^k}(x^i,\mu^i,\nu^i) = I_{h^k}(x,\mu,\nu) \ge I_h(x,\mu,\nu) - \varepsilon.$$

The preceding inequality implies that

$$\liminf_{i\to\infty} W_{h^i}(x^i,\mu^i) = \liminf_{i\to\infty} I_{h^i}(x^i,\mu^i) \ge W_h(x,\mu) - \varepsilon.$$

□

3.2 Dynamic programming

In the following we aim to present other types of value functions which occur in the problem of viscosity solutions of the Hamilton–Jacobi–Bellman (HJB) equation associated with the relaxed upper Hamiltonian H^+ defined by

$$\begin{aligned}
H^+(t,x,\rho) = \min_{\mu\in\mathcal{M}^1_+(\mathbb{S})} \max_{\nu\in\mathcal{M}^1_+(\mathbb{Z})} \{\langle \rho, \int_{\mathbb{Z}} [\int_{\mathbb{S}} f(t,x,s,z)\, d\mu(s)]\, d\nu(z)\rangle \\
+ \int_{\mathbb{Z}} [\int_{\mathbb{S}} J(t,x,s,z)\, d\mu(s)]\, d\nu(z)\}
\end{aligned}$$

where the cost function has the form

$$\int_\tau^1 [\int_{\mathbb{Z}} [\int_{\mathbb{S}} J(t, u_{x,\mu,\nu}(t), s, z)\, d\mu_t(s)]\, d\nu_t(z)]\, dt + g(u_{x,\mu,\nu}(1)),$$

where $J : [0,1] \times \mathbb{E} \times \mathbb{S} \times \mathbb{Z} \to \mathbb{R}$ is an L^1-bounded Carathéodory integrand and $u_{x,\mu,\nu}$ is the trajectory solution starting at position x at intermediate time $\tau \in [0,1[$, associated with the control $(\mu,\nu) \in \mathcal{H} \times \mathcal{R}$, where $\mathcal{H}$ is the set of all Lebesgue-measurable mappings $\mu : [0,1] \to \mathcal{M}^1_+(\mathbb{S})$, namely

$$\dot{u}_{x,\mu,\nu}(t) = \int_{\mathbb{Z}} [\int_{\mathbb{S}} f(t, u_{x,\mu,\nu}(t), s, z)\, d\mu_t(s),]\, d\nu_t(z),\ \ u_{x,\mu,\nu}(\tau) = x$$

and g is an upper-semicontinuous continuous function defined on E (see Remark 1 after Theorem 3.2.1).

In the remainder of this paper, we assume that $\mathcal{H}$ is a *decomposable* subset of the space of Young measures $\mathcal{Y}([0,1];\mathbb{S})$, that is, for every Lebesgue-measurable set $A \subset [0,1]$ and for every μ^1, μ^2 in $\mathcal{H}$, $1_A\mu^1 + 1_{A^c}\mu^2 \in \mathcal{H}$, in particular, $\mathcal{H}$ is the set of all Lebesgue-measurable mappings $\zeta : [0,1] \to S$. Here only decomposability assumption on $\mathcal{H}$ is used by contrast to the results obtained above.

Theorem 3.2.1. (of dynamic programming) *Assume that* $J : [0,1] \times \mathbb{E} \times \mathbb{S} \times \mathbb{Z} \to \mathbb{R}$ *is an* L^1*-bounded Carathéodory integrand. Let* $x \in \mathbb{E}$, *let* $\tau \in [0,1[$ *and let* $\sigma > 0$ *such that* $\tau + \sigma < 1$. *Assume further that* $\mathcal{H}$ *is a decomposable subset of the space of Young measures* $\mathcal{Y}([0,1];\mathbb{S})$. *Let us consider the upper value function*

$$V_J(\tau,x) := \inf_{\mu\in\mathcal{H}} \max_{\nu\in\mathcal{R}}\{\int_\tau^1 [\int_{\mathbb{Z}}[\int_{\mathbb{S}} J(t,u_{x,\mu,\nu}(t),s,z)\,d\mu_t(s)]\,d\nu_t(z)]\,dt\}.$$

Here $u_{x,\mu,\nu}$ *denotes the solution trajectory determined by the relaxed dynamic associated with f and the control* $(\mu,\nu) \in \mathcal{H}\times\mathcal{R}$ *starting at position x at time* $\tau \in [0,1]$, *namely*

$$\dot{u}_{x,\mu,\nu}(t) = \int_{\mathbb{Z}}[\int_{\mathbb{S}} f(t,u_{x,\mu,\nu}(t),s,z)\,\mu_t(s)]\,d\nu_t(z);\; u_{x,\mu,\nu}(\tau) = x.$$

Then the following hold:

(3.2.1)

$$\begin{aligned} V_J(\tau,x) = \inf_{\mu\in\mathcal{H}} \max_{\nu\in\mathcal{R}}\{&\int_\tau^{\tau+\sigma} [\int_{\mathbb{Z}}[\int_{\mathbb{S}} J(t,u_{x,\mu,\nu}(t),s,z)\,d\mu_t(s)]\,d\nu_t(z)]\,dt \\ &+ V_J(\tau+\sigma,u_{x,\mu,\nu}(\tau+\sigma))\}. \end{aligned}$$

Here

$$\begin{aligned} &V_J(\tau+\sigma,u_{x,\mu,\nu}(\tau+\sigma)) \\ &:= \inf_{\beta\in\mathcal{H}} \max_{\gamma\in\mathcal{R}}\{\int_{\tau+\sigma}^1 [\int_{\mathbb{Z}}[\int_{\mathbb{S}} J(t,v_{x,\beta,\gamma}(t),y,z)\,d\beta_t(s)]\,d\gamma_t(z)]\,dt\}. \end{aligned}$$

$v_{x,\beta,\gamma}$ *denotes the trajectory on* $[\tau+\sigma,1]$ *associated with* $(\beta,\gamma) \in \mathcal{H}\times\mathcal{R}$ *with the initial condition* $v_{x,\beta,\gamma}(\tau+\sigma) = u_{x,\mu,\nu}(\tau+\sigma)$.

Proof. We will use the continuity results obtained above. Let $W_J(\tau,x)$ denote the right hand side of (3.2.1). Take $\varepsilon > 0$ and $\mu^1 \in \mathcal{H}$ such that

$$W_J(\tau, x) \geq \max_{\nu \in \mathcal{R}} \{ \int_\tau^{\tau+\sigma} [\int_{\mathbb{Z}} [\int_{\mathbb{S}} J(t, u_{x,\mu^1,\nu}(t), s, z)\, d\mu_t^1(s)]\, d\nu_t(z)]\, dt$$

$$+ V_J(\tau+\sigma, u_{x,\mu^1,\nu}(\tau+\sigma)) \} - \varepsilon.$$

There is $\mu^2 \in \mathcal{H}$ such that

$$V_J(\tau+\sigma, u_{x,\mu^1,\nu}(\tau+\sigma))$$
$$> \max_{\gamma \in \mathcal{R}} \int_{\tau+\sigma}^1 [\int_{\mathbb{Z}} [\int_{\mathbb{S}} J(t, v_{x,\mu^2,\gamma}(t), s, z)\, d\mu_t^2(s)]\, d\gamma_t(z)] dt - \varepsilon.$$

Here $v_{x,\mu^2,\gamma}$ denotes the solution trajectory on $[\tau+\sigma, 1]$ associated with $(\mu^2, \gamma) \in \mathcal{H} \times \mathcal{R}$ with the initial condition $v_{x,\mu^2,\gamma}(\tau+\sigma) = u_{x,\mu^1,\nu}(\tau+\sigma)$. By compactness of $\mathcal{R}$ and by the continuity of

$$\nu \mapsto \int_\tau^{\tau+\sigma} [\int_{\mathbb{Z}} [\int_{\mathbb{S}} J(t, u_{x,\mu,\nu}(t), s, z)\, d\mu_t(s)]\, d\nu_t(z)]\, dt$$

on $\mathcal{R}$, we can choose $\nu^1 \in \mathcal{R}$ such that

$$\max_{\nu \in \mathcal{R}} \int_\tau^{\tau+\sigma} [\int_{\mathbb{Z}} [\int_{\mathbb{S}} J(t, u_{x,\mu^1,\nu}(t), s, z)\, d\mu_t^1(s)]\, d\nu_t(z)]\, dt$$
$$= \int_\tau^{\tau+\sigma} [\int_{\mathbb{Z}} [\int_{\mathbb{S}} J(t, u_{x,\mu^1,\nu^1}(t), s, z)\, d\mu_t^1(s)]\, d\nu_t^1(z)]\, dt$$

and similarly there is $\nu^2 \in \mathcal{R}$ such that

$$\max_{\gamma \in \mathcal{R}} \int_{\tau+\sigma}^1 [\int_{\mathbb{Z}} [\int_{\mathbb{S}} J(t, v_{x,\mu^2,\gamma}(t), s, z)\, d\mu_t^2(s)]\, d\gamma_t(z)]\, dt$$
$$= \int_{\tau+\sigma}^1 [\int_{\mathbb{Z}} [\int_{\mathbb{S}} J(t, v_{x,\mu^2,\nu^2}(t), s, z)\, \mu_t^2(s)]\, d\nu_t^2(z)]\, dt.$$

Here v_{x,μ^2,ν^2} is the trajectory solution defined on $[\tau+\sigma, 1]$ associated with $(\mu^2, \nu^2) \in \mathcal{H} \times \mathcal{R}$ with the initial condition $v_{x,\mu^2,\nu^2}(\tau+\sigma) = u_{x,\mu^1,\nu}(\tau+\sigma)$. Let us set

$$\overline{\mu} := 1_{[\tau,\tau+\sigma]}\mu^1 + 1_{[\tau+\sigma,1]}\mu^2$$

and

$$\overline{\nu} := 1_{[\tau,\sigma]}\nu^1 + 1_{[\tau+\sigma,1]}\nu^2.$$

By the decomposability of $\mathcal{H}$ and $\mathcal{R}$ (recall that $\mathcal{R}$ is the set of measurable selections of a multifunction), we have that $\overline{\mu} \in \mathcal{H}$ and $\overline{\nu} \in \mathcal{R}$. Let $\overline{w}_{x,\overline{\mu},\overline{\nu}}$ be the trajectory on $[\tau, 1]$ associated with $(\overline{\mu}, \overline{\nu}) \in \mathcal{H} \times \mathcal{R}$ with the initial condition

x, that is, $\overline{w}_{x,\overline{\mu},\overline{\nu}}(t) = u_{x,\mu^1,\nu^1}(t)$ for $t \in [\tau, \tau+\sigma]$ and $\overline{w}_{x,\overline{\mu},\overline{\nu}}(t) = v_{x,\mu^2,\nu^2}(t)$ for $t \in [\tau+\sigma, 1]$. Coming back to the definition of $V_J(\tau, x)$, we have

$$\begin{aligned} V_J(\tau,x) &\leq \sup_{\nu\in\mathcal{R}} \int_\tau^1 [\int_{\mathbb{Z}} [\int_{\mathbb{S}} J(t, u_{x,\overline{\mu},\nu}, s, z)\, d\overline{\mu}_t(s)]\, d\nu_t(z)]\, dt \\ &= \int_\tau^{\tau+\sigma} [\int_{\mathbb{Z}} [\int_{\mathbb{S}} J(t, u_{x,\mu^1,\nu^1}, s, z)\, d\mu_t^1(s)]\, d\nu_t^1(z)]\, dt \\ &\quad + \int_{\tau+\sigma}^1 [\int_{\mathbb{Z}} [\int_{\mathbb{S}} J(t, v_{x,\mu^2,\nu^2}, s, z)\, d\mu_t^2(s)]\, d\nu_t^2(z)]\, dt \\ &= \int_\tau^1 [\int_{\mathbb{Z}} [\int_{\mathbb{S}} J(t, \overline{w}_{x,\overline{\mu},\overline{\nu}}, s, z)\, d\overline{\mu}_t(s)]\, d\overline{\nu}_t(z)]\, dt \\ &\leq W_J(\tau,x) + 2\varepsilon. \end{aligned}$$

On the other hand, there is $\widetilde{\mu} \in \mathcal{H}$ such that

$$V_J(\tau,x) \geq \max_{\nu\in\mathcal{R}}\{\int_\tau^1 [\int_{\mathbb{Z}} [\int_{\mathbb{S}} J(t, u_{x,\widetilde{\mu},\nu}(t), s, z)\, \widetilde{\mu}_t(s)]\, d\nu_t(z)]\, dt\} - \varepsilon.$$

Then

$$\begin{aligned} W_J(\tau,x) \leq \max_{\nu\in\mathcal{R}}\{\int_\tau^{\tau+\sigma} & [\int_{\mathbb{Z}} [\int_{\mathbb{S}} J(t, u_{x,\widetilde{\mu},\nu}(t), s, z)\, d\widetilde{\mu}_t(s)]\, d\nu_t(z)]\, dt \\ & + V_J(\tau+\sigma, u_{x,\widetilde{\mu},\nu}(\tau+\sigma))\}, \end{aligned}$$

with

$$\begin{aligned} & V_J(\tau+\sigma, u_{x,\widetilde{\mu},\nu}(\tau+\sigma)) \\ & \leq \max_{\nu\in\mathcal{R}} \int_{\tau+\sigma}^1 [\int_{\mathbb{Z}} [\int_{\mathbb{S}} J(t, u_{x,\widetilde{\mu},\nu}(t), s, z)\, d\widetilde{\mu}_t(s)]\, d\nu_t(z)] dt. \end{aligned}$$

Here $v_{x,\widetilde{\mu},\nu}$ is the trajectory solution associated with $(\widetilde{\mu}, \nu) \in \mathcal{H} \times \mathcal{R}$ with the initial condition $v_{x,\widetilde{\mu},\nu}(\tau+\sigma) = u_{x,\widetilde{\mu},\nu}(\tau+\sigma)$. By continuity of the integral functionals under consideration and by compactness of $\mathcal{R}$, there exists $\overline{\nu^1} \in \mathcal{R}$ such that

$$\begin{aligned} & \max_{\nu\in\mathcal{R}} \int_\tau^{\tau+\sigma} [\int_{\mathbb{Z}} [\int_{\mathbb{S}} J(t, u_{x,\widetilde{\mu},\nu}, s, z)\, d\widetilde{\mu}_t(s)]\, d\nu_t(z)]\, dt \\ & \qquad = \int_\tau^{\tau+\sigma} [\int_{\mathbb{Z}} [\int_{\mathbb{S}} J(t, u_{x,\widetilde{\mu},\overline{\nu^1}}, s, z)\, d\widetilde{\mu}_t(s)]\, d\overline{\nu^1}_t(z)]\, dt \end{aligned}$$

and similarly there exists $\overline{\nu}^2 \in \mathcal{R}$ such that

$$\int_{\tau+\sigma}^{1} J(t, v_{x,\widetilde{\mu},\overline{\nu}^2}(t), s, z)\, \widetilde{\mu}_t(s)]\, d\overline{\nu}_t^2(z)]\, dt$$

$$= \max_{\nu \in \mathcal{R}} \int_{\tau+\sigma}^{1} J(t, v_{x,\widetilde{\mu},\nu}(t), s, z)\, d\widetilde{\mu}_t(s)]\, d\nu_t(z)]\, dt.$$

Here $v_{x,\widetilde{\mu},\overline{\nu}^2}$ is the trajectory on $[\tau+\sigma, 1]$ associated with $(\widetilde{\mu}, \overline{\nu}^2) \in \mathcal{H} \times \mathcal{R}$ with the initial condition $v_{x,\widetilde{\mu},\overline{\nu}^2}(\tau+\sigma) = u_{x,\widetilde{\mu},\nu}(\tau+\sigma)$. Let $\widetilde{\nu} = 1_{[\tau,\tau+\sigma]}\overline{\nu^1} + 1_{[\tau+\sigma,1]}\overline{\nu}^2 \in \mathcal{R}$ and $\widetilde{u}_{x,\widetilde{\mu},\widetilde{\nu}}(t) = u_{x,\widetilde{\mu},\nu^1}(t)$ if $t \in [\tau, \tau+\sigma]$ and $\widetilde{u}_{x,\widetilde{\mu},\widetilde{\nu}}(t) = v_{x,\widetilde{\mu},\overline{\nu}^2}(t)$ if $t \in [\tau+\sigma, 1]$. Taking the above estimate of $W_J(\tau, x)$ into account, we get

$$\begin{aligned} W_J(\tau, x) &\leq \int_\tau^1 [\int_{\mathbb{Z}} [\int_{\mathbb{S}} J(t, \widetilde{u}_{x,\widetilde{\mu},\widetilde{\nu}}(t), s, z)\, \widetilde{\mu}_t(s)]\, d\widetilde{\nu}_t(z)]\, dt \\ &\leq \max_{\nu \in \mathcal{R}} \int_\tau^1 [\int_{\mathbb{Z}} [\int_{\mathbb{S}} J(t, u_{x,\widetilde{\mu},\nu}(t), s, z)\, d\widetilde{\mu}_t(s)]\, d\nu_t(z)]\, dt \\ &\leq V_J(\tau, x) + \varepsilon. \end{aligned}$$

□

Remarks. 1) The preceding result still holds when the upper value functions has the form

$$\begin{aligned} V_J^g(\tau, x) = \inf_{\mu \in \mathcal{H}} \max_{\nu \in \mathcal{R}} \{ \int_\tau^1 [\int_{\mathbb{Z}} [\int_{\mathbb{S}} J(t, u_{x,\mu,\nu}(t), y, z)\, d\mu_t(s)]\, d\nu_t(z)]\, dt \\ + g(u_{x,\mu,\nu}(1)) \}, \end{aligned}$$

where g is a bounded upper semicontinuous function defined on $\mathbb{E}$. Here the decomposability properties of $\mathcal{H}$ and $\mathcal{R}$ and the compactness of $\mathcal{R}$ for the stable topology in the space of Young measures $\mathcal{Y}([0,1]; \mathbb{Z})$ are very useful in the proof of Theorem 3.2.1. Similarly, one can give an analogous result for the lower value function

$$U_J^g(\tau, x)$$

$$= \sup_{\mathcal{R}} \inf_{\mathcal{H}} \{ \int_\tau^{\tau+\sigma} [\int_{\mathbb{Z}} [\int_{\mathbb{S}} J(t, u_{x,\mu,\nu}(t), s, z)\, d\mu_t(s)]\, d\nu_t(z)]\, dt + g(u_{x,\mu,\nu}(1)) \}$$

by permuting the role of $\mathcal{R}$ and $\mathcal{H}$ and by assuming that g is a bounded lower semicontinous function defined on $\mathbb{E}$. 2) It is worth to mention that the preceding result holds in the particular case where $\mathcal{H}$ is the set of all Lebesgue-measurable mappings from $[0,1]$ to $\mathcal{M}_+^1(\mathbb{S})$ endowed with the vague topology $\sigma(\mathcal{C}(\mathbb{S})^*, \mathcal{C}(\mathbb{S}))$.

In the sequel of this paper, we will make the following assumptions.

(H_1) $\mathcal{H} = \mathcal{Y}([0,1];\mathbb{S})$ and $\mathcal{K} = \mathcal{Y}([0,1];\mathbb{Z})$. In particular, the sets $\mathcal{H}$ and $\mathcal{K}$ are decomposable and compact for the stable topology.

(H_2) $f : [0,1] \times \mathbb{E} \times \mathbb{S} \times \mathbb{Z} \to \mathbb{E}$ is bounded continuous, f is uniformly Lipschitz with respect to the variable $x \in \mathbb{E}$ and the family $(f(.,.,s,z))_{(s,z)\in\mathbb{S}\times\mathbb{Z}}$ is equicontinuous; $J : [0,1] \times \mathbb{E} \times \mathbb{S} \times \mathbb{Z} \to \mathbb{R}$ is bounded continuous and the family $(J(.,.,s,z))_{(s,z)\in\mathbb{S}\times\mathbb{Z}}$ is equicontinuous.

It is obvious that the mappings $\widetilde{f} : [0,1] \times \mathbb{E} \times \mathcal{M}^1_+(\mathbb{S}) \times \mathcal{M}^1_+(\mathbb{Z}) \to \mathbb{E}$ and $\widetilde{J} : [0,1] \times \mathbb{E} \times \mathcal{M}^1_+(\mathbb{S}) \times \mathcal{M}^1_+(\mathbb{Z}) \to \mathbb{R}$ defined by

$$\widetilde{f}(t,x,\mu,\nu) = \int_{\mathbb{Z}} [\int_{\mathbb{S}} f(t,x,s,z)\, d\mu(s)]\, d\nu(z)$$

and

$$\widetilde{J}(t,x,\mu,\nu) = \int_{\mathbb{Z}} [\int_{\mathbb{S}} J(t,x,s,z)\, d\mu(s)]\, d\nu(z)$$

inherit the properties of f and J respectively.

The remainder is a careful adaptation of the techniques given in [El187] and [ES84]. Before going further let us mention a useful lemma.

Lemma 3.2.2. *Let $(t_0, x_0) \in [0,1] \times \mathbb{E}$, and let $\Lambda : [t_0,1] \times \mathbb{E} \times \mathcal{M}^1_+(\mathbb{S}) \times \mathcal{M}^1_+(\mathbb{Z}) \to \mathbb{R}$ be a continuous mapping such that the family $(\Lambda(.,.,\mu,\nu))_{(\mu,\nu)\in\mathcal{M}^1_+(\mathbb{S})\times\mathcal{M}^1_+(\mathbb{Z})}$ is equicontinuous.*

(a) *If $\min_{\mu\in\mathcal{M}^1_+(\mathbb{S})} \max_{\nu\in\mathcal{M}^1_+(\mathbb{Z})} \Lambda(t_0,x_0,\mu,\nu) < -\eta < 0$ for some $\eta > 0$, then there exist $\overline{\mu} \in \mathcal{H}$ and $\sigma > 0$ such that*

$$\max_{\nu\in\mathcal{K}} \int_{t_0}^{t_0+\sigma} \Lambda(t, u_{x_0,\overline{\mu},\nu}(t), \overline{\mu}, \nu_t)\, dt < -\frac{\sigma\eta}{2},$$

where $u_{x_0,\overline{\mu},\nu}$ denotes the trajectory solution of the relaxed dynamic associated with f and the controls $\overline{\mu}$ and ν with the initial condition $u_{x_0,\overline{\mu}_t,\nu}(t_0) = x_0$, that is,

$$\dot{u}_{x_0,\overline{\mu},\nu}(t) = \int_{\mathbb{Z}} [\int_{\mathbb{S}} f(t, u_{x,\overline{\mu},\nu}(t), s, z)\, d\overline{\mu}_t(s)]\, d\nu_t(z) \ a.e.\ t \in [0,1],$$
$$u_{x_0,\overline{\mu},\nu}(t_0) = x_0.$$

(b) *If $\min_{\mu\in\mathcal{M}^1_+(\mathbb{S})} \max_{\nu\in\mathcal{M}^1_+(\mathbb{Z})} \Lambda(t_0,x_0,\mu,\nu) > \eta > 0$ for some $\eta > 0$, then there exists $\sigma > 0$ such that, for each $\mu \in \mathcal{H}$, we have*

$$\max_{\nu\in\mathcal{K}} \int_{t_0}^{t_0+\sigma} \Lambda(t, u_{x_0,\mu,\nu}(t), \mu_t, \nu_t)\, dt > \frac{\sigma\eta}{2}.$$

Proof. (a) By hypothesis, there is $\overline{\mu} \in \mathcal{M}^1_+(\mathbb{S})$ such that $\max_{\nu\in\mathcal{M}^1_+(\mathbb{Z})} \Lambda(t_0, x_0, \overline{\mu}, \nu) < -\eta < 0$. Also, by the equicontinuity hypothesis, there exists $\xi > 0$ such that $\max_{\nu\in\mathcal{M}^1_+(\mathbb{Z})}\Lambda(t, x, \overline{\mu}, \nu) < -\eta/2$ for $0 \le t - t_0 \le \xi$ and $\|x - x_0\| \le \xi$. Take $\sigma > 0$ such that $\int_{t_0}^{t_0+\sigma} c(t)\,dt \le \xi$ so that $\|u_{x_0,\overline{\mu},\nu}(t) - u_{x_0,\overline{\mu},\nu}(t_0)\| \le \int_{t_0}^{t_0+\sigma} c(t)\,dt \le \xi$ for all $t \in [t_0, t_0 + \sigma]$ and for all $\nu \in \mathcal{K}$ (we also denote by $\overline{\mu}$ the constant Young measure $t \mapsto \overline{\mu}_t = \overline{\mu}$). Then, by integrating,

$$\int_{t_0}^{t_0+\sigma} \Lambda(t, u_{x_0,\overline{\mu},\nu}(t), \overline{\mu}, \nu_t)\,dt \le \int_{t_0}^{t_0+\sigma} \max_{\nu'\in\mathcal{M}^1_+(\mathbb{Z})} \Lambda(t, u_{x_0,\overline{\mu},\nu}(t), \overline{\mu}, \nu')\,dt$$
$$< -\frac{\sigma\eta}{2}$$

for all $\nu \in \mathcal{K}$ and the result follows.

(b) Let $\xi > 0$ such that, for all $(\mu, \nu) \in \mathcal{M}^1_+(\mathbb{S}) \times \mathcal{M}^1_+(\mathbb{Z})$, if $0 \le t - t_0 < \xi$ and $\|x - x_0\| \le \xi$, then $|\Lambda(t, x, \mu, \nu) - \Lambda(t_0, x_0, \mu, \nu| < \eta/2$. Let $\mu : [0, 1] \to \mathcal{M}^1_+(\mathbb{S})$ be a Lebesgue–measurable mapping. By virtue of von Neumann–Aumann selection theorem (see [CV77, Chapter III], there exists a Lebesgue–measurable mapping $\nu^\mu : [0, 1] \to \mathcal{M}^1_+(\mathbb{Z})$ such that

$$\Lambda(t_0, x_0, \mu_t, \nu^\mu_t) = \max_{\nu\in\mathcal{M}^1_+(\mathbb{Z})} \Lambda(t_0, x_0, \mu_t, \nu)$$

for all $t \in [0, 1]$, because the nonempty compact valued multifunction

$$t \mapsto \{\nu \in \mathcal{M}^1_+(\mathbb{Z});\ \Lambda(t_0, x_0, \mu_t, \nu) = \max_{\nu'\in\mathcal{M}^1_+(\mathbb{Z})} \Lambda(t_0, x_0, \mu_t, \nu')\}$$

has its graph in $\mathcal{L}([0, 1])\otimes\mathcal{B}(\mathcal{M}^1_+(\mathbb{Z}))$. Take $\sigma > 0$ such that $\int_{t_0}^{t_0+\sigma} c(t)\,dt \le \xi$ so that $\|u_{x_0,\mu,\nu}(t) - u_{x_0,\mu,\nu}(t_0)\| \le \int_{t_0}^{t_0+\sigma} c(t)\,dt \le \xi$ for all $t \in [t_0, t_0 + \sigma]$ and for all $\nu \in \mathcal{K}$. As in (a), we then have, by integrating,

$$\int_{t_0}^{t_0+\sigma} \Lambda(t, u_{x_0,\mu,\nu^\mu}(t), \mu_t, \nu^\mu_t)\,dt \ \ge \int_{t_0}^{t_0+\sigma} [\Lambda(t, x_0, \mu_t, \nu^\mu_t) - \frac{\eta}{2}]\,dt$$
$$> \int_{t_0}^{t_0+\sigma} \frac{\eta}{2}\,dt = \frac{\sigma\eta}{2}.$$

□

Theorem 3.2.3. (of viscosity solution) *Assume that* (H_1) *and* (H_2) *are satisfied. Let us consider the lower value function*

$$U_J(\tau, x) := \max_{\nu \in \mathcal{K}} \min_{\mu \in \mathcal{H}} \{ \int_\tau^1 [\int_{\mathbb{Z}} [\int_{\mathbb{S}} J(t, u_{x,\mu,\nu}(t), s, z)\, d\mu_t(s)]\, d\nu_t(z)]\, dt \}.$$

Here $u_{x,\mu,\nu}$ is the trajectory solution to

$$\dot{u}_{x,\mu,\nu}(t) = \int_{\mathbb{Z}} [\int_{\mathbb{S}} f(t, u_{x,\mu,\nu}(t), s, z)\, d\mu_t(s)]\, d\nu_t(z),\ u_{x,\mu,\nu}(\tau) = x.$$

Let us consider the relaxed upper Hamiltonian

$$H^+(t, x, \rho) := \min_{\mu \in \mathcal{M}_+^1(\mathbb{S})} \max_{\nu \in \mathcal{M}_+^1(\mathbb{Z})} \{ \langle \rho, \int_{\mathbb{Z}} [\int_{\mathbb{S}} f(t, x, s, z)\, d\mu(s)]\, d\nu(z) \rangle + \int_{\mathbb{Z}} [\int_{\mathbb{S}} J(t, x, s, z)\, d\mu(s)]\, d\nu(z) \}.$$

Then U_J is a viscosity solution to the HJB equation $\frac{\partial U}{\partial t} + H^+(t, x, \nabla U) = 0$, that is, for any $\varphi \in \mathcal{C}^1([0,1] \times \mathbb{E})$ for which $U_J - \varphi$ attains a local maximum at (t_0, x_0), we have

$$\frac{\partial \varphi}{\partial t}(t_0, x_0) + H^+(t_0, x_0, \nabla\varphi(t_0, x_0)) \geq 0,$$

and for any $\varphi \in \mathcal{C}^1([0,1]) \times \mathbb{E})$ for which $U_J - \varphi$ attains a local minimum at (t_0, x_0), we have

$$\frac{\partial \varphi}{\partial t}(t_0, x_0) + H^+(t_0, x_0, \nabla\varphi(t_0, x_0)) \leq 0.$$

Proof. Assume by contradiction that there exists a $\varphi \in \mathcal{C}^1([0,1] \times \mathbb{E})$ and a point (t_0, x_0) for which

$$\frac{\partial \varphi}{\partial t}(t_0, x_0) + H^+(t_0, x_0, \nabla\varphi(t_0, x_0)) \leq -\eta < 0 \text{ for } \eta > 0.$$

Applying Lemma 3.2.2-(a), by taking $\Lambda = \widetilde{J} + \langle \nabla\varphi, \widetilde{f} \rangle + \frac{\partial \varphi}{\partial t}$, where, for all $(t, x, \mu, \nu) \in [0,1] \times \mathbb{E} \times \mathcal{M}_+^1(\mathbb{S}) \times \mathcal{M}_+^1(\mathbb{Z})$,

$$\widetilde{J}(t, x, \mu, \nu) = \int_{\mathbb{Z}} [\int_{\mathbb{S}} J(t, x, s, z)\, d\mu(s)]\, d\nu(z)$$

and

$$\widetilde{f}(t, x, \mu, \nu) = \int_{\mathbb{Z}} [\int_{\mathbb{S}} f(t, x, s, z)\, d\mu(s)]\, d\nu(z),$$

provides a control $\overline{\mu} \in \mathcal{M}_+^1(\mathbb{S})$ and $\sigma > 0$ such that

$$\max_{\nu\in\mathcal{K}}\{\int_{t_0}^{t_0+\sigma}[\int_{\mathbb{Z}}[\int_{\mathbb{S}} J(t,u_{x_0,\overline{\mu},\nu}(t),s,z)\,d\overline{\mu}(s)]\,d\nu_t(z)]\,dt$$
$$+\int_{t_0}^{t_0+\sigma}[\int_{\mathbb{Z}}[\int_{\mathbb{S}}\langle\nabla\varphi(t,u_{x_0,\overline{\mu},\nu}(t)),f(t,u_{x_0,\overline{\mu},\nu}(t),s,z)\rangle\,d\overline{\mu}(s)]\,d\nu_t(z)]\,dt$$
$$+\int_{t_0}^{t_0+\sigma}\frac{\partial\varphi}{\partial t}(t,u_{x_0,\overline{\mu},\nu}(t))\,dt\}$$
$$\leq -\sigma\eta/2.$$

Thus

$$\max_{\nu\in\mathcal{K}}\min_{\mu\in\mathcal{H}}\{\int_{t_0}^{t_0+\sigma}[\int_{\mathbb{Z}}[\int_{\mathbb{S}} J(t,u_{x_0,\mu,\nu}(t),s,z)\,d\mu(s)]\,d\nu_t(z)]\,dt$$
$$+\int_{t_0}^{t_0+\sigma}[\int_{\mathbb{Z}}[\int_{\mathbb{S}}\langle\nabla\varphi(t,u_{x_0,\mu,\nu}(t)),f(t,u_{x_0,\mu,\nu}(t),s,z)\rangle\,d\mu_t(s)]\,d\nu_t(z)]\,dt$$
$$+\int_{t_0}^{t_0+\sigma}\frac{\partial\varphi}{\partial t}(t,u_{x_0,\mu,\nu}(t))\,dt\}$$
$$(3.2.2)\ \leq -\sigma\eta/2.$$

From the dynamic programming identity (see Remark 1 following Theorem 3.2.1),

$$U_J(t_0,x_0)=\max_{\nu\in\mathcal{K}}\min_{\mu\in\mathcal{H}}\{\int_{t_0}^{t_0+\sigma}[\int_{\mathbb{Z}}[\int_{\mathbb{S}} J(t,u_{x,\mu,\nu}(t),s,z)\,d\mu_t(s)]\,d\nu_t(z)]\,dt$$
$$+U_J(t_0+\sigma,u_{x_0,\mu,\nu}(t_0+\sigma))\}. \tag{3.2.3}$$

Here $u_{x_0,\mu,\nu}$ is the trajectory solution of the relaxed dynamic $\widetilde{f}$ associated with $\mu\in\mathcal{H}$ and $\nu\in\mathcal{K}$ with the initial condition $u_{x_0,\zeta,\nu}(t_0)=x_0$. Since $U_J-\varphi$ has a local maximum at (t_0,x_0), so for small enough σ
(3.2.4)
$$U_J(t_0,x_0)-\varphi(t_0,x_0)\geq U_J(t_0+\sigma,u_{x_0,\mu,\nu}(t_0+\sigma))-\varphi(t_0+\sigma,u_{x_0,\mu,\nu}(t_0+\sigma)).$$

From (3.2.3) and (3.2.4)

$$\max_{\nu\in\mathcal{K}}\min_{\mu\in\mathcal{H}}\{\int_{t_0}^{t_0+\sigma}[\int_{\mathbb{Z}}[\int_{\mathbb{S}} J(t,u_{x,\mu,\nu}(t),s,z)\,d\mu_t(s)]\,d\nu_t(z)]\,dt$$
$$+\varphi(t_0+\sigma,u_{x_0,\mu,\nu}(t_0+\sigma)-\varphi(t_0,x_0))\} \tag{3.2.5}$$
$$\geq 0.$$

As φ is $\mathcal{C}^1$

$$
\begin{aligned}
&\varphi(t_0+\sigma, u_{x_0,\mu,\nu}(t_0+\sigma)) - \varphi(t_0, x_0) \\
&= \int_{t_0}^{t_0+\sigma} \langle \nabla\varphi(t, u_{x_0,\mu,\nu}(t)), \\
&\quad [\int_{\mathbb{Z}} [\int_{\mathbb{S}} f(t, u_{x_0,\mu,\nu}(t), s, z)\, d\mu_t(s)]\, d\nu_t(z)]\,\rangle dt \\
&\quad + \int_{t_0}^{t_0+\sigma} \frac{\partial\varphi}{\partial t}(t, u_{x_0,\mu,\nu}(t))\, dt.
\end{aligned}
\tag{3.2.6}
$$

Substituting (3.2.6) in (3.2.5) we have a contradiction to (3.2.2). Therefore we must have

$$
\frac{\partial\varphi}{\partial t}(t_0, x_0) + H^+(t_0, x_0, \nabla\varphi(t_0, x_0)) \geq 0.
$$

The verification for the min point is similar, using Lemma 3.2.2-(b). □

Now we give a variant of the dynamic programming identity with strategies. Let $\mathcal{A} \subset \mathcal{Y}([0,1];\mathbb{S})$ be compact for the convergence in probability (note that $\mathcal{A}$ is then compact for the stable topology) and let $\mathcal{B} \subset \mathcal{Y}([0,1];\mathbb{Z})$ be compact (metrizable) for the stable topology, e.g. $\mathcal{B} = \mathcal{R}$. We denote by Δ the set of all continuous mappings (strategies) $\alpha : \mathcal{A} \to \mathcal{B}$. Then the dynamic programming identity becomes as follows.

Theorem 3.2.4. The dynamic programming identity with strategies *Let $x \in \mathbb{E}$ and let $\sigma, \tau \in [0,1[$ such that $\tau+\sigma < 1$. Let $J : [0,1]\times E\times S\times Z \to \mathbb{R}$ be an L^1-bounded Carathéodory integrand. Let us consider the lower value function*

$$
H_J(\tau, x) :=
$$

$$
\inf_{\alpha\in\Delta} \max_{\mu\in\mathcal{A}} \{ \int_\tau^1 [\int_{\mathbb{Z}} [\int_{\mathbb{S}} J(t, u_{x,\mu,\alpha(\mu)}(t), s, y, z)\, d\mu_t(s)]\, d\alpha(\mu)_t(z)]\, dt.\}
$$

Here $u_{x,\mu,\alpha(\mu)}$ denotes the solution trajectory determined by the relaxed dynamic associated with f and the controls $\alpha(\mu)$, ($\alpha \in \Delta$) and $\mu \in \mathcal{A}$ starting at position x at time $\tau \in [0,1[$, that is,

$$
\dot{u}_{x,\mu,\alpha(\mu)}(t) = \int_{\mathbb{Z}} [\int_{\mathbb{S}} f(t, u_{x,\mu,\alpha(\mu)}(t), s, z)\, d\mu_t(s)\, d\alpha(\mu)_t(z)];\ u_{x,\mu,\alpha(\mu)}(\tau) = x.
$$

Then the following holds:

$$H_J(\tau, x) =$$

$$\inf_{\alpha\in\Delta}\max_{\mu\in\mathcal{A}}\{\int_\tau^{\tau+\sigma}[\int_{\mathbb{Z}}[\int_{\mathbb{S}} J(t, u_{x,\mu,\alpha(\mu)}(t), s, z), d\mu_t(s)]\, d\alpha(\mu)_t(z)]\, dt$$

$$(3.2.7) \qquad + H_J(\tau+\sigma, u_{x,\mu,\alpha(\mu),\nu}(\tau+\sigma)\}.$$

Here

$$H_J(\tau+\sigma, u_{x,\mu,\alpha(\mu),}(\tau+\sigma))$$

$$:= \inf_{\gamma\in\Delta}\sup_{\mu\in\mathcal{A}}\{\int_{\tau+\sigma}^1[\int_{\mathbb{Z}}[\int_{\mathbb{S}} J(t, v_{x,\mu,\gamma(\mu)}(t), y, z)\, d\mu_t(s)]\, d\gamma(\mu)_t(z)]\, dt\}$$

where $v_{x,\mu,\gamma(\mu)}$ denotes the trajectory on $[\tau+\sigma, 1]$ associated with $(\gamma(\mu), \mu)$ $(\gamma \in \Delta, \mu \in \mathcal{A})$ with the initial condition $v_{x,\mu,\gamma(\mu)}(\tau+\sigma) = u_{x,\mu,\alpha(\mu)}(\tau+\sigma)$.

Proof. Let $K_J(\tau, x)$ be the second member of (3.2.7). Take $\varepsilon > 0$ and $\gamma^1 \in \Delta$ such that

$$K_J(\tau, x) \geq \max_{\mu\in\mathcal{A}}\{\int_\tau^{\tau+\sigma}[\int_{\mathbb{Z}}[\int_{\mathbb{S}} J(t, u_{x,\mu,\gamma^1(\mu)}(t), s, z)\, d\mu_t(s)]\, d\gamma^1(\mu)_t(z)]\, dt$$

$$+ H_J(\tau+\sigma, u_{x,\mu,\gamma^1(\mu)}(\tau+\sigma)\} - \varepsilon.$$

There is $\gamma^2 \in \Delta$ such that

$$H_J(\tau+\sigma, u_{x,\mu,\gamma^1(\mu)}(\tau+\sigma))$$

$$> \sup_{\mu\in\mathcal{A}}\int_{\tau+\sigma}^1[\int_{\mathbb{Z}}[\int_{\mathbb{S}} J(t, v_{x,\mu,\gamma^2(\mu)}(t), s, z)\, d\mu_t(s)]\, d\gamma^2(\mu)_t(z)]dt - \varepsilon.$$

Here $v_{x,\mu,\gamma^2(\mu)}$ denotes the trajectory of the dynamic on $[\tau+\sigma, 1]$ associated with $(\gamma^2(\mu), \mu)$ $(\gamma^2 \in \Delta, \mu \in \mathcal{A})$ and the initial condition $v_{x,\mu,\gamma^2(\mu)}(\tau+\sigma) = u_{x,\mu,\gamma^1(\mu)}(\tau+\sigma)$. In view of the continuity of γ^1 and the Fiber Product Lemma (Theorem 2.3.1), we see that the mapping

$$\mu \mapsto \int_\tau^{\tau+\sigma}[\int_{\mathbb{Z}}[\int_{\mathbb{S}} J(t, u_{x,\mu,\gamma^1(\mu)}(t), s, z)\, \mu_t(s)]\, d\gamma^1(\mu)_t(z)]\, dt$$

is continuous on the compact (for the convergence in probability) set $\mathcal{A}$. Hence we can choose $\mu^1 \in \mathcal{A}$ such that

$$\max_{\mu\in\mathcal{A}} \int_\tau^{\tau+\sigma} [\int_{\mathbb{Z}}[\int_{\mathbb{S}} J(t, u_{x,\mu,\gamma^1(\mu)}(t), s, z)\, d\mu_t(s)]\, d\gamma^1(\mu)_t(z)]\, dt$$
$$= \int_\tau^{\tau+\sigma} [\int_{\mathbb{Z}}[\int_{\mathbb{S}} J(t, u_{x,\mu^1,\gamma^1(\mu^1)}(t), s, z)\, d\mu_t^1(s)]\, d\gamma^1(\mu^1)_t(z)]\, dt$$

where $u_{x,\mu^1,\gamma^1(\mu^1)}$ is the solution trajectory defined on $[\tau, \tau+\sigma]$ associated with $(\gamma^1(\mu^1), \mu^1)$ $(\gamma^1 \in \Delta, \mu^1 \in \mathcal{A})$. Similarly there is $\mu^2 \in \mathcal{A}$ such that

$$\max_{\mu\in\mathcal{A}} \int_{\tau+\sigma}^{1} [\int_{\mathbb{Z}}[\int_{\mathbb{S}} J(t, v_{x,\mu,\gamma^2(\mu)}(t), s, z)\, d\mu_t(s)]\, d\gamma^2(\mu)_t(z)]\, dt$$
$$= \int_{\tau+\sigma}^{1} [\int_{\mathbb{Z}}[\int_{\mathbb{S}} J(t, v_{x,\mu^2,\gamma^2(\mu^2)}(t), s, z)\, d\mu_t^2(s)]\, d\gamma^2(\mu^2)_t(z)]\, dt.$$

Let us define $\overline{\gamma} \in \Delta$ by setting, for all $\mu \in \mathcal{A}$: $\overline{\gamma}(\mu)_t = \gamma^1(\mu)_t$ for $t \in [\tau, \sigma]$ and $\overline{\gamma}(\mu)_t = \gamma^2(\mu)_t$ for $t \in [\tau+\sigma, 1]$. Coming back to the definition of $H_J(\tau, x)$

$$H_J(\tau, x) \le \sup_{\mu\in\mathcal{A}} \int_\tau^1 [\int_{\mathbb{Z}}[\int_{\mathbb{S}} J(t, u_{x,\mu,\overline{\gamma}(\mu)}, s, z)\, d\mu_t(s)]\, d\overline{\gamma}(\mu)_t(z)]\, dt$$
$$= \int_\tau^{\tau+\sigma} [\int_{\mathbb{Z}}[\int_{\mathbb{S}} J(t, u_{x,\mu^1,\gamma^1(\mu^1)}, s, z)\, d\mu_t^1(s)]\, d\gamma^1(\mu^1)_t(z)]\, dt$$
$$+ \int_{\tau+\sigma}^{1} [\int_{\mathbb{Z}}[\int_{\mathbb{S}} J(t, v_{x,\mu^2,\gamma^2(\mu)}, s, z)\, d\mu_t^2(s)]\, d\gamma^2(\mu^2)_t(z)]\, dt$$
$$\le K_J(\tau, x) + 2\varepsilon.$$

On the other hand, there is $\widetilde{\gamma} \in \Delta$ such that

$$H_J(\tau, x) \ge \sup_{\mu\in\mathcal{A}} \{\int_\tau^1 [\int_{\mathbb{Z}}[\int_{\mathbb{S}} J(t, u_{x,\mu,\widetilde{\gamma}(\mu)}(t), s, z)\, d\mu_t(s)]\, d\widetilde{\gamma}(\mu)_t(z)]\, dt\} - \varepsilon.$$

Then

$$K_J(\tau, x) \le \max_{\mu\in\mathcal{A}} \{ \int_\tau^{\tau+\sigma} [\int_{\mathbb{Z}}[\int_{\mathbb{S}} J(t, u_{x,\mu,\widetilde{\gamma}(\mu)}(t), s, z)\, d\mu_t(s)]\, d\widetilde{\gamma}(\mu)_t(z)]\, dt$$
$$+ H_J(\tau+\sigma, u_{x,\mu,\widetilde{\gamma}(\mu)}(\tau+\sigma)\}$$

with

$$H_J(\tau+\sigma, u_{x,\mu,\widetilde{\gamma}(\mu)}(\tau+\sigma)$$
$$\le \sup_{\mu\in\mathcal{A}} \int_{\tau+\sigma}^{1} [\int_{\mathbb{Z}}[\int_{\mathbb{S}} J(t, v_{x,\mu,\widetilde{\gamma}(\mu)}(t), s, z)\, \mu_t(s)]\, d\widetilde{\gamma}(\mu)_t(z)]dt.$$

Here $v_{x,\mu,\widetilde{\gamma}}(\mu)$ is the trajectory on $[\tau+\sigma,1]$ associated with $(\widetilde{\gamma}(\mu^2),\mu^2)$ with the initial condition $v_{x,\mu,\widetilde{\gamma}(\mu)}(\tau+\sigma) = u_{x,\mu,\widetilde{\gamma}(\mu)}(\tau+\sigma)$. As above, by continuity of the integral functionals under consideration and by compactness of $\mathcal{A}$, there exists $\mu^1 \in \mathcal{A}$ such that

$$\max_{\mu\in\mathcal{A}} \int_\tau^{\tau+\sigma} [\int_{\mathbb{Z}}[\int_{\mathbb{S}} J(t,u_{x,\mu,\widetilde{\gamma}(\mu)},s,z)\,\mu_t(s)]\,d\widetilde{\gamma}(\mu)_t(z)]\,dt$$
$$= \int_\tau^{\tau+\sigma} [\int_{\mathbb{Z}}[\int_{\mathbb{S}} J(t,u_{x,\mu^1,\widetilde{\gamma}(\mu^1)},s,z)\,d\mu^1_t(s)]\,d\widetilde{\gamma}(\mu^1)_t(z)]\,dt.$$

Here $u_{x,\widetilde{\gamma}(\overline{\mu^1}),\overline{\mu^1}}$ is the trajectory on $[\tau,\tau+\sigma]$ associated with $(\widetilde{\gamma}(\overline{\mu^1}),\overline{\mu^1})$ and similarly there is $\mu^2 \in \mathcal{A}$ such that

$$\int_{\tau+\sigma}^{1} [\int_{\mathbb{Z}}[\int_{\mathbb{S}} J(t,v_{x,\mu^2,\widetilde{\gamma}(\mu^2)}(t),s,z)\,d\widetilde{\gamma}(\overline{\mu}^2)_t(s)]\,d\widetilde{\gamma}(\mu^2)_t(z)]\,dt$$
$$= \max_{\mu\in\mathcal{A}} \int_{\tau+\sigma}^{1} [\int_{\mathbb{Z}}[\int_{\mathbb{S}} J(t,v_{x,\mu,\widetilde{\gamma}(\mu)}(t),s,z)\,d\widetilde{\mu}_t(s)]\,d\gamma(\mu)_t(z)]\,dt.$$

Let

$$w_{x,\widetilde{\gamma}(\mu),\mu}(t) = \begin{cases} u_{x,\widetilde{\gamma}(\overline{\mu^1}),\overline{\mu^1}}(t) & \text{for } t\in[\tau,\tau+\sigma] \\ v_{x,\widetilde{\gamma}(\mu^2),\overline{\mu}^2}(t) & \text{for } t\in[\tau+\sigma,1]. \end{cases}$$

Taking the expression of $K_J(\tau,x)$ into account, we get

$$K_J(\tau,x) \le \int_\tau^1 [\int_{\mathbb{Z}}[\int_{\mathbb{S}} J(t,w_{x,\widetilde{\gamma}(\mu),\mu}(t),s,z)\,d\widetilde{\gamma}(\mu)_t(s)]\,d\mu_t(z)]\,dt$$
$$\le \sup_{\mu\in\mathcal{A}} \int_\tau^1 [\int_{\mathbb{Z}}[\int_{\mathbb{S}} J(t,u_{x,\widetilde{\gamma}(\mu),\mu}(t),s,z)\,d\widetilde{\gamma}(\mu)_t(s)]\,d\mu_t(z)]\,dt$$
$$\le H_J(\tau,x)+\varepsilon.$$

□

Remarks. The preceding techniques can be applied to the case when $\mathcal{A}$ is equal to $\mathcal{M}^1_+(\mathbb{S})$, Γ is the set of all Carathéodory mappings $\alpha : [0,1]\times\mathcal{M}^1_+(\mathbb{S}) \to \mathcal{M}^1_+(\mathbb{Z})$ and Δ is the set of all mappings $\hat{\alpha} : \mathcal{M}^1_+(\mathbb{S}) \to \mathcal{Y}([0,1],\mathbb{Z})$ given by $\hat{\alpha}(\mu)_t = \alpha_t(\mu)$ for all $\alpha\in\Gamma$, $\mu\in\mathcal{M}^1_+(\mathbb{S})$ and for all $t\in[0,1]$.

Aknowledgements We heartily thank Michel Valadier and Lionel Thibault for helpful discussions and comments. We remain responsible for all shortcomings in this paper.

References

[Bal84a] Balder, E.J.: A general approach to lower semicontinuity and lower closure in optimal control theory, SIAM J. control and Optimization **22** 570–598 (1984)

[Bal84b] Balder, E.: A general denseness result for relaxed control theory, Bull. Austral. Math. Soc. **30** 463–475 (1984)

[Bal88] ———: Generalized equilibrium results for games with incomplete information, Math. Oper. Res. **13** no. 2, 265–276 (1988)

[Bal89] ———: On Prohorov's theorem for transition probabilities, Séminaire d'Analyse Convexe, **19** 9.1-9.11 (1989)

[Bal90] ———: New sequential compactness results for spaces of scalarly integrable functions, J. Math. Anal. Appl. **151** 1–16 (1990)

[Bal00] ———: New fundamentals of Young measure convergence, pp. 24–48, Calculus of Variations and Optimal Control (Haifa 1998) (Boca Raton, FL), Chapman & Hall, 2000

[BJ91] Barron, E.N., Jensen R.: Optimal Control and semicontinuous viscosity solutions, Proc. A. M. S, **113** (2) 397-402 (1991)

[Bil68] Billingsley, P.: Convergence of probability measures, J. Wiley, New York, London, 1968

[Cas80] Castaing, C.: Topologie de la convergence uniforme sur les parties uniformément intégrables de L^1_E et théorèmes de compacité faible dans certains espaces du type Köthe-Orlicz, Séminaire d'Analyse Convexe **10** 5.1-5.27 (1980)

[CJ03] Castaing, C., Jofre, A.: Optimal control and variational problems, Preprint Université de Montpellier, 18 pages 2003

[CV77] Castaing, C., Valadier, M.: Convex Analysis and Measurable multifunctions, Lectures Notes in Math. **580**, Springer-Verlag, Berlin (1977)

[CV98] ———: Weak convergence using Young measures, Funct. Approximatio Comment. Math. **26** 7–17 (1998)

[Dud66] Dudley, R. M.: Convergence of Baire measures, Studia Math. **27** 251–268 (1966)

[EK72] Robert, J., Elliott, Nigel, J. Kalton: The existence of value in differential games, American Mathematical Society, Providence, R.I., Memoirs of the American Mathematical Society, No. 126, 1972

[Ell87] Robert, J. Elliott: Viscosity solutions and optimal control, Pitman Research Notes in Mathematics Series, vol. 165, Longman Scientific & Technical, Harlow, 1987

[ES84] Evans, L. C., Souganidis, P. E.: Differential games and representation formulas for solutions of Hamilton-Jacobi-Isaacs equations, Indiana Univ. Math. J. **33** no. 5, 773–797 (1984)

[Fis70] Fischler, R.: Suites de bi–probabilités stables, Annales de la Faculté des Sciences de l'Université de Clermont **43** 159–167 (1970)

[HJ91] Hoffmann-Jørgensen, J.: Stochastic processes on Polish spaces, Various Publication Series, no. 39, Matematisk Institut, Aarhus Universitet, Aarhus, Denmark, 1991

[HJ98] ———: Convergence in law of random elements and random sets, pp. 151–189, High dimensional probability (Oberwolfach, 1996), Progress in Probability, no. 43, Birkhäuser, Basel, 1998

[Jaw84] Jawhar, A.: Mesures de transition et applications, Séminaire d'Analyse Convexe **13** 13.1–13.62 (1984)

[KS88] Krasovskii, N.N., Subbotin, A.I.: Game-Theoretical Control Problems, Springer-Verlag in Soviet Mathematics, New York, 1988

[RdF03] Raynaud de Fitte, P.: Compactness criteria in the stable topology, Bull. Pol. Acad. Sci. Math. **51**(4) 343-363 (2003)

[Tat02] Tateishi, H.: On the existence of equilibria of equicontinuous games with incomplete information, Adv. Math. Econ. **4** 41–59 (2002)

[Val73] Valadier, M.: Désintégration d'une mesure sur un produit, C. R. Acad. Sci. Paris Sér. I **276** A33–A35 (1973)

[Val90] ———: Young measures, pp. 152–158, Methods of Nonconvex Analysis (Berlin) (A. Cellina, ed.), Lecture Notes in Math., no. 1446, Springer Verlag, 1990

[Val94] ———: A course on Young measures, Rendiconti dell'istituto di matematica dell'Università di Trieste **26, suppl.** 349–394, (1994), Workshop di Teoria della Misura et Analisi Reale Grado, 1993 (Italia)

Adv. Math. Econ. 6, 39–53 (2004)

Advances in
MATHEMATICAL
ECONOMICS

The compactness of $Pr(K)$

Dionysius Glycopantis[1] **and Allan Muir**[2]

[1] Department of Economics, City University, Northampton Square, London EC1V 0HB, UK
(e-mail: d.glycopantis@city.ac.uk)
[2] Department of Mathematics, City University, Northampton Square, London EC1V 0HB, UK
(e-mail: muirlynn@llanfynydd.u-net.com)

Received: April 4, 2003
Revised: May 20, 2003

JEL classification: C72

Mathematics Subject Classification (2000): 91A10

Summary. We prove the compactness of $Pr(K)$, the set of Borel probability measures on a compactum K endowed with the weak* topology, without embedding this set in $rca(K)$, the space of regular, countably-additive, signed measures with their finite total variation as norm. $Pr(K)$ can be extended to a convex, Hausdorff, linear topological space. Then Glicksberg's fixed point theorem is applied to prove the existence of Nash equilibria.

Key words: Borel probability measures, weak* topology, compactness, payoff functions, reaction correspondences, Nash equilibrium, fixed point theorem.

1. Introduction

It is known that the classical theorems of [Nash(1950)] and [Debreu(1952)] can be generalized to show that, under appropriate conditions, in games with a finite number of players and ∞-dimensional pure strategy spaces there exists a Nash equilibrium in mixed strategies, which can be taken to be the Borel probability measures over pure strategies. An original result in the literature is that of [Glicksberg(1952)], obtained through the application of a generalization of the Kakutani fixed point theorem which he proves first. Glicksberg states this generalization thus:

> Given a closed point to convex set mapping $\Phi : S \to S$ of a convex compact subset S of a convex Hausdorff linear topological space into itself there exists a fixed point $x \in \Phi(x)$.

Note that in Glicksberg's definitions a "closed" mapping is one whose graph is closed; moreover, $\Phi(x)$ is required to be non-empty for each $x \in S$.

For this application we need to establish in the first instance the compactness, in the weak* topology which is discussed below, of the set of probability measures, $Pr(K)$, on the Borel sets of a compact Hausdorff space K. The space K is identified with a player's set of pure strategies and $Pr(K)$ with the mixed strategies. One then argues that an appropriate mapping, the reaction correspondences, from the convex compact set of the product of the players' mixed strategies into itself, can be constructed. This has a fixed point which implies the existence of a Nash equilibrium. The convexity of $Pr(K)$ which implies the convexity of the product of players' mixed strategies is obvious but compactness is not, and a justification is needed for the choice of the weak* topology.

Glicksberg does not actually prove the weak* compactness of $Pr(K)$. On the other hand his statement strongly suggests a line of proof which we shall refer to as the *indirect method* for proving compactness.

Briefly, one can treat the set of probability measures on K as embedded in the space of all regular, countably additive, signed measures on the Borel subsets with norm their finite total variation on K. The Riesz representation theorem states that this Banach space is isomorphic to the space of continuous linear functionals, with the supremum norm, defined on the continuous functions on K. One invokes Alaoglu's theorem that the norm-unit sphere in the linear functionals space is compact in the weak* topology and then proves that the set of probability measures is a closed subset, and hence is compact, in this topology.

We go on to discuss an alternative proof which establishes the compactness of $Pr(K)$ more directly. We begin and work throughout the compactness argument with $Pr(K)$ endowed with the weak* topology, cast in a more convenient form than the usual one. We sought such a proof because probability measures are simple in structure but the other elements of the space in which they are embedded are more complicated and more difficult to grasp, so that the mathematical results of the indirect method are unnecessarily powerful. Our approach borrows from the techniques of the theorems of Alaoglu and Riesz without using their full power. Having established the compactness of $Pr(K)$ we can easily extend this space to a convex Hausdorff linear topological space in preparation for the application of Glicksberg's fixed point theorem.

Before we proceed, we clarify a couple of important points. Consider the indirect method of proof. The question arises as to why one needs to go to another topology different than the one defined through the supremum norm.

The reason is that in this topology the unit sphere is in general not compact and therefore we cannot argue that the closed subset of probability measures is also compact, which is needed by the fixed point theorem. The fact that the unit sphere is not always compact follows from the theorem discussed in [Dunford and Schwartz(1957), p. 245] that a normed linear space is finite dimensional if and only if its closed unit sphere is compact. Glicksberg's generalization takes place in infinite dimensional spaces.

We draw attention to the fact that [Billingsley(1968), p.7] discusses regular measures, which are defined below, in the context of metric spaces. Also, [Pryce(1973), pp.217–225] , in proving the Riesz representation theorem, states that he simplifies matters by considering metric spaces. In such spaces closedness of a set can be established simply by examining convergent sequences rather than nets. Furthermore on a compact metric space the σ-algebras of Borel and Baire sets coincide and the regularity property of the elements of $Pr(K)$ is established in a direct manner. However the fixed point theorem of Glicksberg is in the context of a convex Hausdorff linear topological space which might not be metrizable and the generalization of the discussion from metric to Hausdorff spaces requires more complicated arguments.

Urysohn's metrization theorem, discussed in [Dunford and Schwartz(1957), p.24] , states that a compact Hausdorff space is metrizable if and only if its topology has a countable base and this provides a criterion for checking metrizability. The following is an example of a reasonable Hausdorff space for the application of Glicksberg's fixed point theorem which is not metrizable. The space is $\mathbb{R}^I$ with the product topology, where $I = [0, 1]$, and the compact subset is I^I. This subset is not metrizable because it does not have a countable base.

[Royden(1988), p. 238] states in an exercise that the weak* topology on the unit sphere in the dual of a separable Banach space is metrizable. However, as it is pointed out there, this does not guarantee that the whole of the dual space endowed with the weak* topology is itself metrizable. Furthermore even metrizability of the unit sphere in the dual space is not guaranteed for the spaces allowed in Glicksberg's theorem. For example the space of continuous functions on $[0, 1]$ with the supremum norm is a separable Banach space but this is not true for every space consisting of continuous functions defined on a compactum.

The paper is organized as follows: Section 2 contains the notation used and some background concepts and results. In Section 3 we give details of the indirect method and in Section 4 we present our alternative approach for proving the compactness of $Pr(K)$. In Section 5 $Pr(K)$ is extended to a convex Hausdorff linear topological space and the conditions on the reaction functions for the application of Glicksberg's fixed point theorem are proved; hence the existence of a Nash equilibrium is established. The paper ends with some concluding remarks.

2. Notation and preliminaries

A^c denotes the complement of a set A; $A \backslash B$ denotes the elements of A which are not in B.

K will denote a *compactum* — i.e. a compact Hausdorff space. $\mathcal{G}, \mathcal{F}$ denote the families of open sets and closed sets of K, respectively.

$\mathcal{B}(K)$ denotes the σ-algebra of *Borel sets* of K, the smallest σ-algebra containing $\mathcal{G}$.

$\mathcal{B}'(K)$ denotes the σ-algebra of *Baire sets* of K, the smallest σ-algebra with respect to which every real-valued continuous function on K is measurable.

$\mathcal{C}(K)$ denotes the Banach space of real-valued, continuous functions on K, with $x \in \mathcal{C}(K)$ having supremum norm $\|x\| = \max_{t \in K} |x(t)|$.

$\mathcal{C}^+(K)$ denotes the set of positive elements of $\mathcal{C}(K)$,

$$\{x \in \mathcal{C}(K) : \forall t \in K,\ x(t) \geq 0\}.$$

$\mathcal{C}(K)^*$ denotes the dual Banach space of continuous linear functionals on $\mathcal{C}(K)$, with $f \in \mathcal{C}(K)^*$ having supremum norm $\|f\| = \sup_{\|x\|=1} |f(x)|$.

$f \in \mathcal{C}(K)^*$ is said to be *positive* if $x \in \mathcal{C}^+(K)$ implies $f(x) \geq 0$.

Let m be a real-valued function on $\mathcal{B}(K)$, and $A \in \mathcal{B}(K)$. For any finite partition
$\mathcal{P} = \{K_i \in \mathcal{B}(K) : i = 1, \ldots, n\}$ of K form the sum

$$\mathcal{S}(m, A, \mathcal{P}) = \sum_{i=1}^{n} |m(A \cap K_i)|;$$

the supremum of $\mathcal{S}(m, A, \mathcal{P})$, taken over all such finite partitions $\mathcal{P}$, is called *the total variation of* m *over* A, denoted $|m|(A)$.

m is said to be *regular* if for all $A \in \mathcal{B}(K), \varepsilon > 0$, there exist $G \in \mathcal{G}, F \in \mathcal{F}$ with $F \subseteq A \subseteq G$ such that $|m|(G \backslash F) < \varepsilon$.

$rca(K)$ denotes the set of all signed measures — i.e. countably additive, regular functions, m on $\mathcal{B}(K)$ with finite *total variation* $|m|(K)$. In this case $\|m\|_r = |m|(K)$ is called the *norm* of m, which makes $rca(K)$ into a Banach space.

$Pr(K)$ denotes the set of *probability measures* on K, the subset of $rca(K)$ given by

$$\{\mu \in rca(K) : \mu(A) \geq 0 \text{ for all } A \in \mathcal{B}(K); \|\mu\|_r = 1\}.$$

Note that, for such positive measures, $\|\mu\|_r = \mu(K)$.

$S^* = \{f \in \mathcal{C}(K)^* : \|f\| \leq 1\}, S_r^* = \{m \in rca(K) : \|m\|_r \leq 1\}$ denote the unit balls in $\mathcal{C}(K)^*$, $rca(K)$ respectively.

Borel sets and Baire sets are compared in [Royden(1988)] from which the following statements are taken. If X is a compact space, every Baire set is a Borel set. The converse is true when X is a compact *metric* space. However there are compact spaces in which the class of Borel sets is larger than the class of Baire sets. In our case K is compact but not necessarily a metric space, so we must distinguish between Baire and Borel σ-algebras and measures. On the other hand the compactness of K implies that Baire measures can be extended uniquely to Borel measures so that a Borel set differs from some Baire set by a set of measure zero.

Further definitions will be given and crucial theorems invoked as they are needed.

3. $Pr(K)$ as a subset of $\mathcal{C}(K)^*$

In this section we explain how one can prove the compactness of $Pr(K)$ following the indirect method implied in the statements of Glicksberg. We invoke the Riesz representation theorem, proved for example in [Dunford and Schwartz(1957), p. 265], which states that there is a Banach space isomorphism ϕ from $(rca(K), \| \ \|_r)$ to $(\mathcal{C}(K)^*, \| \ \|)$, taking $m \in rca(K)$ to $f \in \mathcal{C}(K)^*$ given by

$$f(x) = \int_K x \, dm \text{ for all } x \in \mathcal{C}(K). \tag{1}$$

Using this we can transfer discussion easily between the two spaces.

Our objective is to argue that $Pr(K)$ is compact because it is a closed subset of the unit ball. However, when we examine the norm topologies we see that from the point of view of applying Glicksberg's fixed point theorem there is a problem. Namely, as explained in the Introduction, we do not necessarily have compactness. Therefore we impose a different topology, the weak* topology, which is the smallest topology that makes the elements of $\mathcal{C}(K)$ continuous on $\mathcal{C}(K)^*$, with the action of $x \in \mathcal{C}(K)$ on $f \in \mathcal{C}(K)^*$ defined through $x(f) = f(x)$. In this topology a basic open neighbourhood of $g \in \mathcal{C}(K)^*$ is a set of the form

$$\{f \in \mathcal{C}(K)^* : |f(x_i) - g(x_i)| < \varepsilon_i\}$$

for some finite set $\{x_1, \ldots, x_n \in \mathcal{C}(K); \ \varepsilon_1, \ldots, \varepsilon_n > 0\}$, which of course is convex.

Using the Riesz mapping defined in equation (1), we can recast the weak* topology in the context of $rca(K)$; this automatically makes ϕ be a homeomorphism between the spaces. A basic open neighbourhood of $m_0 \in rca(K)$ in the weak* topology will be a set of the form

$$\left\{m \in rca(K) : \left|\int_K x_i \, dm - \int_K x_i \, dm_0\right| < \varepsilon_i\right\}$$

for some finite set $\{x_1, \ldots, x_n \in \mathcal{C}(K);\ \varepsilon_1, \ldots, \varepsilon_n > 0\}$. Note that the weak* topology on $rca(K)$ is equivalent to the so-called *topology of weak convergence of measures*. In the latter, convergence of the net $\{m^\delta : \delta \in D\}$ to m, where D is a directed set, is defined by

$$\left\{\int_K x \, dm^\delta : \delta \in D\right\} \to \int_K x \, dm$$

for all $x \in \mathcal{C}(K)$.

We then invoke Alaoglu's theorem, given for example in [Royden(1988), p.237] , that the norm unit sphere S^* is compact in the weak* topology. This implies that the corresponding unit sphere, S_r^*, of $rca(K)$ is compact.

The set in $\mathcal{C}(K)^*$ corresponding under ϕ to $Pr(K)$ is

$$S^{*\prime} = \{f \in S^* : f(x) \geq 0 \text{ for all } x \in \mathcal{C}^+(K); \|f\| = 1\}.$$

This follows, for example, from the discussion in [Kingman and Taylor(1966), p.251] that positive linear functionals can be associated with finite Baire measures, the fact that these can be extended uniquely to Borel measures and from the fact that for probability measures the norm is equal to 1.

We prove that $S^{*\prime}$ is closed in two steps.

Step 1 For given $x \in \mathcal{C}^+(K)$, $\{f \in S^* : f(x) \geq 0\}$ is closed since it is the inverse image of $\{y \in \mathbb{R} : y \geq 0\}$ under the evaluation map $x : \mathcal{C}(K)^* \to \mathbb{R}$, given by $x(f) = f(x)$, which is continuous in the weak* topology. Thus

$$\hat{S} = \{f \in S^* : f(x) \geq 0 \text{ for all } x \in \mathcal{C}^+(K)\} = \bigcap_{x \in \mathcal{C}^+(K)} \{f \in S^* : f(x) \geq 0\}$$

is closed since it is an intersection of closed sets.

Step 2 Note first that for $f \in \hat{S}$ we have $\|f\| = f(\mathbf{1})$, where $\mathbf{1}$ denotes the constant function $\mathbf{1}(t) = 1$ for all $t \in K$. This is because, if $\|x\| = 1$ for $x \in \mathcal{C}(K)$ then $x(t) \leq 1$ for all $t \in K$, then $\mathbf{1} - x \in \mathcal{C}^+(K)$; thus $f(\mathbf{1}) - f(x) = f(\mathbf{1} - x) \geq 0$.

Then $S^{*\prime}$ is the inverse image in $\hat{S}$ of $\{1\}$ under $\mathbf{1} : \mathcal{C}(K)^* \to \mathbb{R}$ and is therefore closed.

Before concluding this section we consider the particular case of a finite number of pure strategies $K = \{k_1, \ldots, k_n\}$. Illustrating with the case $n = 3$, the elements of K can be identified with the three elementary unit vectors in $\mathbb{R}^3$. In order for K to be a Hausdorff space we must endow it with the discrete topology. The Borel σ-algebra is then the power set $\mathcal{P}(K)$ and $rca(K)$ will be the set of countably additive functions, m on $\mathcal{P}(K)$; any such function will be completely determined by its values on k_i and its norm will be $\| m \|_r = \sum_{i=1}^{3} |m(\{k_i\})|$. This norm topology makes $rca(K)$ equivalent to $\mathbb{R}^3$ taken with the Euclidean metric and k_i appears as the point measure with $m(\{k_i\}) = 1$. The discrete topology on K now appears as the subspace topology induced from $rca(K)$ on the point measures. The set $Pr(K)$ consists of the convex combinations of the representation of the pure strategies and it is obviously compact.

$\mathcal{C}(K)$, the space of continuous, real-valued functions x on K, will consist of all vectors $\{x_1, x_2, x_3\}$, where $x_i = x(k_i)$, with $\| x \| = \sup |x_i|$ as norm. The continuous linear functionals, f, on $\mathcal{C}(K)$, i.e. $f \in \mathcal{C}(K)^*$ will be of the form $f(x) = \alpha_1 x_1 + \alpha_2 x_2 + \alpha_3 x_3$ where α_i is a constant, and $\| f \| = \sup_{\|x\|=1} |f(x)| = \sum_{i=1}^{3} |\alpha_i|$. Therefore $\mathcal{C}(K)^*$ is identified with $rca(K)$ and $Pr(K)$ with the compact subset of positive linear functionals, i.e. $\alpha_i \geq 0$ with $\sum_{i=1}^{3} |\alpha_i| = 1$.

It can also be seen, for example from [Royden(1988), p. 234, p.236] , that the weak* topology on $\mathcal{C}(K)^*$, in which a sub-basis element around 0 is given by $\{(\alpha_1, \alpha_2, \alpha_3) : |\sum \alpha_i x_i| < \varepsilon\}$ for a vector $x \in \mathcal{C}(K)$ and $\varepsilon > 0$, is equivalent to the norm topology. It is in the infinite dimensional case that the two topologies do not coincide and we need to go to the weak* topology in order to secure compactness of $Pr(K)$ and apply Glicksberg's fixed point theorem.

4. An alternative approach to compactness

The main aim in this section is to provide a proof of the compactness of the set $Pr(K)$ which focuses directly on $Pr(K)$, not embedding it in the Banach space $rca(K)$ with the rather complicated norm $\| m \|_r$ given in Section 2. Hence we do not get involved with the intricacies of this space, such as the decomposition of signed measures into the difference of two positive measures, the need to establish that the condition of finite total variation holds, or proving the norm closure of a set of measures through the convergence of the supremum of sums of absolute values on the partitions of K, and the isomorphism of the space to $\mathcal{C}(K)^*$ with the supremum norm. We do not have a norm on $Pr(K)$ and at no stage is any space endowed with a norm topology. Instead we impose immediately on $Pr(K)$ the weak* topology and prove its compactness.

Obviously the mathematical arguments and techniques we employ come at some stage close to those one would use in showing the compactness of

$Pr(K)$ through the indirect method but, as we shall see, the full power of these mathematical results is not needed.

In particular although we employ arguments of a similar nature to those in the proof of Alaoglu's theorem, as given for example in [Royden(1988), p. 237], we are able to restrict ourselves to $Pr(K)$. Furthermore, instead of the full Riesz representation theorem we employ the result in [Royden(1988), p.352] concerning positive linear functionals. We shall give details below.

By the weak* topology on $Pr(K)$ we shall understand the one with convex basic open neighbourhoods around μ_0 of the form:

$$\left\{\mu \in Pr(K) : \left|\int_K x_i \, d\mu - \int_K x_i \, d\mu_0\right| < \varepsilon_i\right\}$$

for some finite set $\{x_1, \ldots, x_n \in \mathcal{C}^+(K);\ \varepsilon_1, \ldots, \varepsilon_n > 0\}$. This is the same topology as the one induced on $Pr(K)$ as a subset of $rca(K)$ with weak* topology, since any $x \in \mathcal{C}(K)$ may be written as a difference of elements from $\mathcal{C}^+(K)$. Casting the weak* topology in terms of $x_1, \ldots, x_n \in \mathcal{C}^+(K)$ rather than by anchoring on $x_1, \ldots, x_n \in \mathcal{C}(K)$ is precisely what makes the proof of compactness of $Pr(K)$ easier.

We now prove the main result in this paper:

Theorem 4.1. *$Pr(K)$ is compact in the weak* topology.*

We shall prove the theorem through a series of lemmas.

We start by considering a fixed positive continuous function on K namely $x \in \mathcal{C}^+(K)$ and use it to define a real-valued function on $Pr(K)$

$$\mu \mapsto \int_K x \, d\mu.$$

Since $\int_K x \, d\mu \leq \parallel x \parallel = \max_{t \in K} x(t)$, the image set of this function lies in the interval $[0, \parallel x \parallel]$ and every μ corresponds to a point in the product set

$$D = \prod_{x \in \mathcal{C}^+(K)} [0, \parallel x \parallel].$$

Thus, by Tychonoff's Theorem, the aggregate of all these functions, expressed by

$$\phi : Pr(K) \to D$$

with

$$\phi(\mu) = \left(\int_K x \, d\mu\right)_{x \in \mathcal{C}^+(K)}, \tag{2}$$

has compact codomain D in the product topology.

We note that this exactly parallels the construction at the start of the proof of Alaoglu's theorem, but we are able to confine attention to positive values. In order to follow, for our purposes, the logic of that proof we need to establish

1. ϕ is a homeomorphism to its image set, which means:
   a) ϕ is 1-1.
   b) ϕ establishes a bijective correspondence between the open sets of the weak* topology on $Pr(K)$ and those of the product topology on $im(\phi)$ in D.

   This will imply that both ϕ and ϕ^{-1} are continuous functions.
2. $im(\phi)$ is closed and hence compact. This will imply that $Pr(K)$ is also compact as the image of a compact set under a continuous function. As in the proof of Alaoglu's theorem we shall begin with $\xi \in \overline{im(\phi)}$. The objective is to show that $\xi \in im(\phi)$, i.e. that there exists $\bar{\mu} \in Pr(K)$ such that $\xi(x) = \int_K x \, d\bar{\mu}$.

First we prove:

Lemma 4.2. *$\mu \in Pr(K)$ is regular.*

Proof. *We use the fact that a Borel measure can be restricted to a Baire measure and the fact that, given the compactness of K, the latter can be uniquely extended back to the Borel measure we started with.*

[Royden(1988), p. 337] defines a Baire measure μ to be regular if for each Baire set $E \in \mathcal{B}'(K)$ it is true that $\mu(E)$ is the supremum of the measures of certain compact sets contained in E. From the discussion there it follows that each Baire measure on a compact space is regular. It is then shown in Theorem 11, p. 314, that a regular Baire measure on a compact Hausdorff space can be extended uniquely to a Borel measure and that for each Borel set $E \in \mathcal{B}(K)$ the Borel measure $\mu(E)$ is the supremum of the measures of the compact sets contained in E.

Now a compact set in a Hausdorff space is closed and the above imply that for all $A \in \mathcal{B}(K)$ and $\varepsilon \geq 0$ there is a closed set $F \in \mathcal{F}$ such that $F \subseteq A$ and $\mu(A \setminus F) \leq \varepsilon$. It similarly follows that since $A^c \in \mathcal{B}(K)$ there exists $G \in \mathcal{G}$ such that $G^c \subseteq A^c$ and $\mu(A^c \setminus G^c) = \mu(G \setminus A) \leq \varepsilon$.

Finally from the fact that $(A \setminus F) \cup (G \setminus A) = (G \setminus F)$ and the discussion above follows that for all $A \in \mathcal{B}(K)$ and $\varepsilon \geq 0$ there are $F \in \mathcal{F}$ and $G \in \mathcal{G}$ with $F \subseteq A \subseteq G$ and $\mu(G \setminus F) \leq \varepsilon$ for each Borel set on a compactum K.

This completes the proof of Lemma 4.2.

Of course for $\mu \in Pr(K)$ we have $|\mu| = \mu$. [Billingsley(1968), p.7] shows that for metric spaces the elements of $Pr(K)$ have this property and we proved above that this is true for compact Hausdorff spaces.

Referring to the definition in equation (2)

Lemma 4.3. *ϕ is 1-1.*

Proof. *Consider $\mu \neq \mu'$ both elements of $Pr(K)$. Then there exists $A \in \mathcal{B}(K)$ such that $\mu(A) \neq \mu'(A)$. It follows that for any $\varepsilon > 0$ there are $F,\ F' \in \mathcal{F}$ and $G,\ G' \in \mathcal{G}$ such that $F \subseteq A \subseteq G$ and $F' \subseteq A \subseteq G'$ with $\mu(G \setminus F) < \varepsilon$ and $\mu'(G' \setminus F') < \varepsilon$. Setting $\hat{F} = F \cup F'$ and $\hat{G} = G \cap G'$ we obtain $\hat{F} \subseteq A \subseteq \hat{G}$ with $\mu(\hat{G} \setminus \hat{F}) < \varepsilon$ and $\mu'(\hat{G} \setminus \hat{F}) < \varepsilon$.*

Now since K is compact and Hausdorff it is also a normal space. Therefore by Urysohn's lemma (See [Royden (1988), p.179]) we can choose $x \in \mathcal{C}^+(K)$ such that

$$\begin{aligned} x(t) &= 1 \ \ \text{if} \ \ t \in \hat{F} \\ &= 0 \ \ \text{if} \ \ t \notin \hat{G} \\ &\in [0,1] \ \ \forall\, t \in K. \end{aligned}$$

Applying this function on μ we obtain

$$\int_K x\, d\mu = \int_A x\, d\mu + \int_{\hat{G}\setminus A} x\, d\mu = \mu(A) + \int_{\hat{G}\setminus A} x\, d\mu - \int_{A\setminus \hat{F}} [1-x] d\mu. \quad (3)$$

With a similar result for μ' we can write

$$\begin{aligned} \int_K x\, d\mu - \int_K x\, d\mu' &= \mu(A) - \mu'(A) \\ &+ \int_{\hat{G}\setminus A} x\, d\mu - \int_{\hat{G}\setminus A} x\, d\mu' \\ &+ \int_{A\setminus \hat{F}} [1-x]\, d\mu - \int_{A\setminus \hat{F}} [1-x]\, d\mu'. \end{aligned}$$

All integrals on the right-hand-side have modulus less than ε and therefore

$$\left| \int_K x\, d\mu - \int_K x\, d\mu' \right| \geq |\mu(A) - \mu'(A)| - 4\varepsilon.$$

In particular choosing $\varepsilon = \frac{1}{8}|\mu(A) - \mu'(A)|$ gives

$$\int_K x\, d\mu \neq \int_K x\, d\mu' \Longrightarrow \phi(\mu) \neq \phi(\mu').$$

This completes the proof of Lemma 4.3.
We now prove:

Lemma 4.4. *ϕ is a homeomorphism onto its image.*

Proof. *We know that a basic neighbourhood of $\mu_0 \in Pr(K)$ in the weak* topology can be cast in the form*

$$\left\{\mu \in Pr(K) : \left|\int_K x_i \, d\mu - \int_K x_i \, d\mu_0\right| < \varepsilon_i\right\}$$

for some finite set $\{x_1, \ldots, x_n \in C^+(K); \varepsilon_1, \ldots, \varepsilon_n > 0\}$.

The collection of images $\xi = (\int_K x \, d\mu)_{x \in C^+(K)}$ of this set, lying in D, is

$$\{\xi \in im(\phi) : |\xi(x_i) - \xi_0(x_i)| < \varepsilon_i\}$$

for the same finite set $\{x_1, \ldots, x_n\} \subseteq C^+(K)$ and the same $\varepsilon_1, \ldots, \varepsilon_n > 0$, which is exactly a basic neighbourhood of $\xi_0 = (\int_K x d\mu_0)_{x \in C^+(K)}$ in the subspace topology on $im(\phi)$ induced from the product topology on D. Therefore open sets map bijectively onto open sets of the subspace topology which proves that when the codomain is restricted to $im(\phi)$ both functions ϕ and ϕ^{-1} are continuous.

This completes the proof of Lemma 4.4.

The next step is to show that $im(\phi)$ is closed, which will imply its compactness. First we note that any $\xi \in im(\phi)$ may be extended to a positive linear function on $C(K)$ since, with $\xi(x) = \int_K x \, d\mu$,

$$\xi(\alpha x + \beta x') = \alpha\xi(x) + \beta\xi(x')$$

for $x, x' \in C(K)$ and $\alpha, \beta \geq 0$, and $\xi(x) \geq 0$ for any $x \in C^+(K)$.

Next we show that elements in the closure of $im(\phi)$ are also positive linear functionals.

Lemma 4.5. *Any $\xi \in \overline{im(\phi)}$ is positively linear on $C^+(K)$.*

Proof. *We argue in a similar manner as at the end of the proof of Alaoglu's theorem in [Royden(1988)] .*

Consider the basic open neighbourhoods of ξ in D generated by x, x', $\alpha x + \beta x' \in C^+(K)$ and all $\varepsilon = \frac{1}{n}$. The fact that ξ is a point of closure means that each such neighbourhood contains some element $\xi_n \in im(\phi)$ i.e.

$$|\xi(x) - \xi_n(x)|, |\xi(x') - \xi_n(x')|, |\xi(\alpha x + \beta x') - \xi_n(\alpha x + \beta x')| < \frac{1}{n}. \quad (4)$$

Setting

$$\begin{aligned} J &= \xi(\alpha x + \beta x') - \alpha\xi(x) - \beta\xi(x') \\ J_n &= \xi_n(\alpha x + \beta x') - \alpha\xi_n(x) - \beta\xi_n(x') \end{aligned}$$

we obtain

$$|J - J_n| < (1+\alpha+\beta)\frac{1}{n}.$$

So $J_n \to J$. However $J_n = 0$, by the linearity of ξ_n, and therefore $J = 0$ which means that ξ is linear. Finally from the fact that $\xi_n(x) \geq 0$, for all $x \in \mathcal{C}^+(K)$, it follows that also $\xi(x) \geq 0$.

This completes the proof of Lemma 4.5.

Now we know that $\xi \in \overline{im(\phi)}$ is positively on $\mathcal{C}^+(K)$. We also know from the Riesz type theorem in [Royden(1988), p. 352] that, since K is compact, for any positive linear functional f on $\mathcal{C}^+(K)$ there is a unique Baire measure $\hat{\mu}$ such that

$$f(x) = \int_K x \, d\hat{\mu} \tag{5}$$

where $x \in \mathcal{C}^+(K)$.

Furthermore we know that this Baire measure can be uniquely extended to a Borel measure and the value of the integral in equation (5) is not affected, as it is independent of the increasing sequence of simple functions chosen and the sequence resting on $\mathcal{B}'(K)$ is also available when $\mathcal{B}(K)$ is considered. Finally the Borel measure $\bar{\mu}$ obtained is a probability measure, i.e., $\bar{\mu}(K) = \hat{\mu}(K) = 1$. This follows from the fact that it is a point of closure of $Pr(K)$ and therefore any of its neighbourhoods contains a probability measure. Choosing neighbourhoods to refer to the constant function $x(t) = 1$ for all t it is seen that $\bar{\mu}(K) = 1$.

Therefore there is a probability measure $\bar{\mu} \in Pr(K)$ which maps onto $\xi \in \overline{im(\phi)}$, so $im(\phi)$ is closed and hence compact. It follows from the continuity of ϕ^{-1} that $Pr(K)$ is compact as it is the image of a compact set under a continuous function.

This completes the proof of the theorem.

5. Nash equilibria

We consider 2-person games, as the generalization to n-player games is obvious. The compacta of pure strategies of the players are denoted by P_i for Player i, with $p_i \in P_i$. Mixed strategies will be denoted by μ_i and they will be elements of $M_i = Pr(P_i)$.

The utility functions of the two players $\hat{u}_1(p_1, p_2)$ and $\hat{u}_2(p_1, p_2)$ are continuous on the compact set $P_1 \times P_2$, in the product topology. The expected utility payoff to Player i is:

$$u_i(\mu_1, \mu_2) = \int_{P_2}\int_{P_1} \hat{u}_i(p_1, p_2) d\mu_1 d\mu_2.$$

The reaction correspondences of the two players are represented by

$$\mu_2 \mapsto \mathcal{R}^1(\mu_2)$$

for Player 1, where $\mathcal{R}^1(\mu_2) = \{\mu_1 : \; u_1(\mu_1, \mu_2) = \max_{\mu'_1 \in M_1} \; u_1(\mu'_1, \mu_2)\}$ gives the set of optimal responses of Player 1 to fixed μ_2 and

$$\mu_1 \mapsto \mathcal{R}^2(\mu_1)$$

is defined *mutatis mutandis* for Player 2.

A Nash equilibrium is defined as a pair of mixed strategies (μ_1^0, μ_2^0) such that

$$\begin{aligned} &\forall \mu_1 \in M_1 : u_1(\mu_1^0, \mu_2^0) \geq u_1(\mu_1, \mu_2^0) \\ \text{and} \quad &\forall \mu_2 \in M_2 : u_2(\mu_1^0, \mu_2^0) \geq u_1(\mu_1^0, \mu_2). \end{aligned}$$

Note that the set of such Nash equilibria is the intersection of the reaction correspondence graphs.

We obtained above the compactness of M_i. Note that for the application of Glicksberg's fixed point theorem, to prove the existence of a Nash equilibrium, we require M_i to be a compact subset of a convex Hausdorff linear topological space. This is achieved by taking all finite linear combinations of elements of M_i, under the equivalence of their action on $\mathcal{B}(P_i)$, and imposing the weak* topology. In the new setting, compactness of M_i is retained since every cover of M_i can be reduced to one in the subspace topology.

We also need to show that $\mathcal{R}^1(\mu_2)$ and $\mathcal{R}^2(\mu_1)$ are both non-empty and convex and that the correspondence

$$(\mu_1, \mu_2) \;\; \mapsto \;\; \mathcal{R}^1(\mu_2) \times \mathcal{R}^2(\mu_1) \tag{6}$$

which is defined on a set with the product of weak* topologies into itself has a closed graph. Then the fixed point theorem will apply so the existence of a fixed point, which in our context will be a Nash equilibrium, is guaranteed.

We concentrate on Player 1. [Glycopantis and Muir(2000)] proves that the expected payoff function

$$u_1(\mu_1, \mu_2) = \int_{P_1} \int_{P_2} \hat{u}_1(p_1, p_2) \; d\mu_2 \; d\mu_1$$

is a continuous function of both variables. So for given $\mu_2 \in M_2, u_1(\mu_1, \mu_2)$ is continuous on the compact set M_1; thus the maximum is attained and $\mathcal{R}^1(\mu_2)$ is non-empty.

Since $u_1(\mu_1, \mu_2)$ is linear in μ_1 for fixed μ_2, for $a, b \geq 0, a + b = 1$

$$\begin{aligned} u_1(\mu_1^1, \mu_2) &\geq u_1(\mu_1, \mu_2) \\ u_1(\mu_1^2, \mu_2) &\geq u_1(\mu_1, \mu_2) \\ \Rightarrow u_1(a\mu_1^1 + b\mu_1^2, \mu_2) &= au_1(\mu_1^1, \mu_2) + bu_1(\mu_1^2, \mu_2) \\ &\geq (a+b)u_1(\mu_1, \mu_2) \\ &= u_1(\mu_1, \mu_2). \end{aligned}$$

So if μ_1^1, μ_1^2 are optimizing for given μ_2, then so is $a\mu_1^1 + b\mu_1^2$; thus $\mathcal{R}^1(\mu_2)$ is convex.

For any $\mu_1' \in M_1$,

$$(\mu_1, \mu_2) \mapsto u_1(\mu_1, \mu_2) - u_1(\mu_1', \mu_2)$$

is continuous, so

$$\{(\mu_1, \mu_2) \in M_1 \times M_2 : u_1(\mu_1, \mu_2) - u_1(\mu_1', \mu_2) \geq 0\}$$

is closed. Then graph of $\mathcal{R}^1$,

$$\begin{aligned} &\{(\mu_1, \mu_2) \in M_1 \times M_2 : \forall \mu_1' \in M_1 : u_1(\mu_1, \mu_2) - u_1(\mu_1', \mu_2) \geq 0\} \\ = &\bigcap_{\mu_1' \in M_1} \{(\mu_1, \mu_2) \in M_1 \times M_2 : u_1(\mu_1, \mu_2) - u_1(\mu_1', \mu_2) \geq 0\}, \end{aligned}$$

is closed.

It follows from the above that the mapping in equation (6) satisfies the conditions of Glicksberg's fixed point theorem and therefore a Nash equilibrium exists.

6. Concluding remarks

The main purpose of the paper was two-fold. First it was to see an explicit proof of the compactness of $Pr(K)$ because there does not appear to be one in the literature and second to provide a proof that is simpler than the one implied there. By a simpler proof we understand one which uses less powerful tools. Also, with respect to proving the existence of a Nash equilibrium, in games with a finite number of players and ∞-dimensional strategy spaces, which is deduced from the fact that the reaction correspondences mapping has a fixed point, we made explicit the proof suggested by Glicksberg. Namely, we provided a proof that the reaction correspondences mapping has a closed graph, which suffices for the application of the fixed point theorem.

References

[Berge(1963)] Berge, C.: Topological spaces. London: Oliver and Boyd 1963

[Billingsley(1968)] Billingsley, P.: Convergence of probability measures. New York London Sydney Toronto: Wiley 1968

[Debreu(1952)] Debreu, C.: A social equilibrium existence theorem. Proc. Natl. Acad. Sci. **38**, 886-893, 1952

[Dunford and Schwartz(1957)] Dunford, N., Schwartz, J.T.: Linear operators, part I. New York: Interscience Publishers, Inc. 1957

[Glicksberg(1952)] Glicksberg, I.L.: A further generalization of the Kakutani Fixed Point Theorem, with applications to Nash equilibrium points. Proceedings of the American Mathematical Society **3**, 170-174, 1952

[Glycopantis and Muir(2000)] Glycopantis, D., Muir, A.: Continuity of the payoff function. Economic Theory, **16**, 239–244, 2000

[Kingman and Taylor(1966)] Kingman, J. F. C., Taylor, S.J.: Introduction to measure and probability. Cambridge: Cambridge University Press 1966

[Nash(1950)] Nash, J.F.: Equilibrium points in N-person games. Proc. Natl. Acad. Sci. **36**, 48-49 1950

[Pryce(1973)] Pryce, J.D.: Basic methods of linear functional analysis. London: Hutchinson 1973

[Royden(1988)] Royden. H.L.: Real Analysis, 3rd edn. London: The Macmillan Company, Collier-Macmillan 1988

Adv. Math. Econ. 6, 55–68 (2004)

Advances in
MATHEMATICAL
ECONOMICS

Recursive methods in probability control

Seiichi Iwamoto

Department of Economic Engineering, Graduate School of Economics, Kyushu University 27, Hakozaki 6-19-1, Higashiku, Fukuoka 812-8581, Japan

Received: April 14,2003
Revised: July 22,2003

JEL classification: C61, D81

Mathematics Subject Classification (2000): 90C39, 90C40, 90A43

Abstract. In this paper we propose recursive methods in probability control. We maximize the threshold probability that a "total" reward is greater than or equal to a given lower level. We take three types of reward function — general function, forward recursive function and backward recursive function. By three types of imbedding method, we derive a backward recursive equation for maximum probability function. The first is to view any history (alternating sequence of state and decision) as a new state. The second is to incorporate past values (cumulative rewards) in state dynamics. The third is a state expansion by attachment of future values (threshold levels). It is shown that the threshold probability problem has an optimal policy in a common large class.

Key words: recursive function, probability control, dynamic programming, invariant imbedding, forward recursive, backward recursive, backward recursive formula, forward recursive formula, past values, cumulative rewards, future values, threshold levels, expanded Markov policy, general policy, primitive policy

1. Introduction

Recently, recursive methods have been extensively applied in economic dynamics (e.g., Ozaki and Streufert [10], Stokey and Lucas [12], Streufert [14] et al.). One of the recursive methods is dynamic programming (Bellman [1]). It is well known that dynamic programming has solved a large class of stochastic optimzation problems. Then, almost all criteria are *expected value* of total reward in a wide sense — additive [3, 4, 13]. The fundamental property of additive criterion is separation under expetation operator. When we adopt *probability* itself as a criterion, what happens? The probability does not separate additive criterion. Here is an optimization-oriented motivation for the study.

On the other hand, probability evaluation is rougher than expected value one. However, it is of practical use. In fact, decision is frequently made on available probabilities at the moment. This is also a motivation.

In general, the threshold probability has been an important criterion in engineering, especially in quality control and reservoir control. Then a total amount/benefit should be subject to a constraint level. As for economics, the probability criterion is not so well studied. However, for instance, a regulation of the gross weight, a correction of the excessive decline in a currency and a supply-demand imbalance will apply the threshold probability that the quantity should be within a given level or between a lower and a upper levels. Some of the expected criteria in stochastic dynamics will be converted into the threshold probability (Le Van and Dana [9], Sargent [11], [12], [14] and others). These are future problems in recursive theory in economic dynamics.

This paper focuses on a threshold probability criterion. We optimize the probability that a total reward function is greater than or equal to a given lower level. It is shown that invariant imbedding is a useful approach to solve the threshold probability problem with recursive structure or without. Through an suitable imbedding [2], the *recursive* threshold problem is reduced to a *terminal* one on an expanded Markov class. We show that the threshold probability problem has an optimal policy in general class.

In section 2 we formulate the threshold probability optimization problem over general class. Section 3 derives an optimal policy in primitive class. Further the optimal policy turns out to generate an optimal policy in general class. In section 4 we maximize the probability for forward recursive function through invariant imbedding. This expands the state space by attaching the cumulative rewards. Section 5 discusses backward recursive function. The stream of remaining threshold levels is incorporated in the original Markov transition law. Finally we give a comment that the methods apply to any probability criterion as well as sandwich probability one.

2. Formulation on general class

Throughout the paper, let $\{(X_n, U_n)\}_0^N$ be an N-stage *controlled Markov chain* on a finite *state space* X and a finite *control space* U with a *Markov transition law* $p = \{p(y|x,u)\}$:

$$p(y|x,u) \ \stackrel{\triangle}{=}\ P(X_{n+1} = y \,|\, X_n = x, U_n = u).$$

This is expressed

$$X_{n+1} \sim p(\cdot \,|\, x_n, u_n) \qquad 0 \le n \le N{-}1.$$

We use the following notations:

$$H_n := X \times U \times X \times U \times \cdots \times X \times U \times X \ ((2n+1)\text{-factors}) \tag{1}$$
$$h_n := (x_0, u_0, x_1, u_1, \ldots, x_n) \in H_n$$
$$X^n := X \times X \times \cdots \times X \ (n\text{-times}).$$

Let any real valued function $g : H_N \to R^1$ and a *lower level* $\underline{c}(\in R^1)$ be given. Then we consider the probability that the *general reward function*

$$G := g(X_0, U_0, X_1, U_1, \ldots, X_{N-1}, U_{N-1}, X_N) \tag{2}$$

is greater than or equal to the lower level $\underline{c}$. We would maximize the *threshold probability* $P(G \geq \underline{c})$ over some policy class, where an optimal policy can exist. Our concern is two-fold. One is recursive structure. The other is how we derive an optimal policy in the class.

Let us now introduce three classes of policies. A *Markov* (resp. *general, primitive*) *policy* $\pi = \{\pi_0, \pi_1, \ldots, \pi_{N-1}\}$ (resp. $\sigma = \{\sigma_0, \sigma_1, \ldots, \sigma_{N-1}\}$, $\mu = \{\mu_0, \mu_1, \ldots, \mu_{N-1}\}$) is a sequence of *Markov* (resp. *general, primitive) decision functions*:

$$\pi_n : X \to U \ \ (\text{resp. } \sigma_n : X^{n+1} \to U, \ \ \mu_n : H_n \to U).$$

The Markov (resp. general, primitive) policy memorizes *current-state* (resp. *past-state*, *past-history*). Let Π (resp. $\Pi(g)$, $\Pi(p)$) denote the set of all Markov (resp. general, primitive) policies. We call Π (resp. $\Pi(g)$, $\Pi(p)$) *Markov* (resp. *general, primitive*) *class*. Hence

$$\Pi \subset \Pi(g) \subset \Pi(p). \tag{3}$$

Further, for n $(0 \leq n \leq N{-}1)$, let Π_n (resp. $\Pi_n(g)$, $\Pi_n(p)$) denote the set of all corresponding policies which start from n-th stage on. For instance, $\Pi_n(p)$ denotes the set of all primitive policies $\mu = \{\mu_n, \ldots, \mu_{N-1}\}$ which begin at stage n.

Let us now formulate the threshold probability optimization problem. We choose general class $\Pi(g)$ as policy class where optimization is taken, because an optimal policy in primitive class need not be in Markov class Π :

$$\mathrm{P}_0(x_0) \qquad \begin{array}{lll} \text{Optimize} & P^{\sigma}_{x_0}(G \geq \underline{c}) & \\ \text{subject to} & (\mathrm{i})_\mathrm{n} \ \ X_{n+1} \sim p(\cdot \mid x_n, u_n) & \\ & (\mathrm{ii})_\mathrm{n} \ \ u_n \in U & \end{array} \quad 0 \leq n \leq N{-}1 \tag{4}$$

where $P^{\sigma}_{x_0}$ is the (discrete) probability measure on the product space X^N induced from the transition law p, a general policy $\sigma = \{\sigma_0, \sigma_1, \ldots, \sigma_{N-1}\} \in \Pi(g)$, and an initial state $x_0 \in X$. Thus we have the partial multiple sum :

$$P^{\sigma}_{x_0}(G \geq \underline{c}) \quad = \quad \sum\sum_{(x_1,\ldots,x_N) \in *}\cdots\sum p_1 p_2 \cdots p_N \tag{5}$$

where $*$ is the set

$$* := \{(x_1, x_2, \ldots, x_N) \,|\, G(h) \geq \underline{c}\} \subset X^N$$

and

$$h = (x_0, u_0, x_1, u_1, \ldots, x_{N-1}, u_{N-1}, x_N)$$
$$p_1 = p(x_1|x_0, u_0), \;\; p_2 = p(x_2|x_1, u_1), \;\; \ldots, \;\; p_N = p(x_N|x_{N-1}, u_{N-1})$$
$$u_0 = \sigma_0(x_0), \;\; u_1 = \sigma_1(x_0, x_1), \;\; \ldots, \;\; u_{N-1} = \sigma_N(x_0, x_1, \ldots, x_{N-1}).$$

For simplicity we write the *threshold probability problem* as follows :

$$\mathrm{P}_0(x_0) \qquad \begin{array}{ll} \text{Optimize} & P^{\sigma}_{x_0}(\,G \geq \underline{c}\,) \\ \text{subject to} & \text{(i)}_\text{n}, \;\; \text{(ii)}_\text{n} \;\; 0 \leq n \leq N\text{–}1. \end{array}$$

Let $v_0(x_0)$ denote the *maximum value* of problem $\mathrm{P}_0(x_0)$:

$$v_0(x_0) \;\; := \;\; \underset{\sigma \in \Pi(g)}{\mathrm{Max}} \; P^{\sigma}_{x_0}(\,G \geq \underline{c}\,) \quad x_0 \in X. \tag{6}$$

Then our problem is to find an *optimal* policy σ^* in general class $\Pi(g)$:

$$v_0(x_0) \;\; = \;\; P^{\sigma^*}_{x_0}(\,G \geq \underline{c}\,) \quad \forall x_0 \in X. \tag{7}$$

3. General criterion

In this section, we consider a *large* family of subproblems $\mathcal{R} = \{\mathrm{R}_n(h_n)\}$:

$$\mathrm{R}_n(h_n) \qquad \begin{array}{lll} \text{Optimize} & P^{\mu}_{h_n}(\,G \geq \underline{c}\,) & \\ \text{subject to} & \text{(i)}_\text{m} \;\; X_{m+1} \sim p(\cdot \,|\, x_m, u_m) & \\ & \text{(ii)}_\text{m} \;\; u_m \in U & n \leq m \leq N\text{–}1 \end{array}$$

$$h_n \in H_n, \;\; 0 \leq n \leq N.$$

The subproblem $\mathrm{R}_n(h_n)$ starts at a given history $h_n \in H_n$ on the n-th stage (see also [5],[6],[7]). The probability measure $P^{\mu}_{h_n}$ is induced from the transition law p, a primitive policy $\mu = \{\mu_n, \ldots, \mu_{N-1}\} \in \Pi_n(p)$, and a history $h_n = (x_0, u_0, \ldots, u_{n-1}, x_n) \in H_n$. We define the maximum value functions $\{w_n\}$ as follows :

$$w_n(h_n) \;\; := \;\; \underset{\mu \in \Pi_n(p)}{\mathrm{Max}} \; P^{\mu}_{h_n}(\,G \geq \underline{c}\,) \;\; h_n \in H_n, \; 0 \leq n \leq N\text{–}1 \tag{8}$$

where

$$w_N(h_N) \quad := \quad 1_{\underline{c}}(g(h_N)) \quad h_N \in H_N. \tag{9}$$

Henceforth $1_d(\cdot)$ denotes the characteristic function of interval $[d, \infty)$:

$$1_d(g) := \begin{cases} 1 & \text{if } g \geq d \\ 0 & \text{otherwise} \end{cases}$$

Then we have the backward recursive relation:

Proposition 1. *(Backward recursive formula)*

$$w_n(h) = \operatorname*{Max}_{u \in U} \sum_{y \in X} w_{n+1}(h, u, y) p(y|x, u) \tag{10}$$

$$h \in H_n, 0 \leq n \leq N\text{–}1$$

$$w_N(h) = 1_{\underline{c}}(g(h)) \quad h \in H_N.$$

Proof. Straightforward.

Thus we have

Theorem 1. *Let $\mu_n^*(h)$ be the set of all maximizers in (11). Then policy $\mu^* = \{\mu_0^*, \mu_1^*, \ldots, \mu_{N-1}^*\}$ is optimal in primitive class; for all $\mu \in \Pi(p)$*

$$P_{x_0}^{\mu^*}(G \geq \underline{c}) \quad \geq \quad P_{x_0}^{\mu}(G \geq \underline{c}) \quad \forall x_0 \in X.$$

In general, for any given $\mu \in \Pi(p)$, we define $\sigma = \{\sigma_0, \sigma_1, \ldots, \sigma_{N-1}\}$ as follows:

$$\begin{aligned} \sigma_0(x_0) \quad &:= \quad \mu_0(x_0) \\ \sigma_1(x_0, x_1) \quad &:= \quad \mu_1(x_0, u_0, x_1) \quad \text{where } u_0 = \mu_0(x_0) \\ &\vdots \\ \sigma_{N-1}(x_0, x_1, \ldots, x_{N-1}) \quad &:= \quad \mu_{N-1}(x_0, u_0, x_1, \ldots, u_{N-2}, x_{N-1}) \end{aligned} \tag{11}$$

$$\begin{aligned} \text{where } u_0 &= \mu_0(x_0), \\ u_1 &= \mu_1(x_0, u_0, x_1), \\ &\ldots, \\ u_{N-2} &= \mu_{N-2}(x_0, u_0, x_1, \ldots, u_{N-3}, x_{N-2}). \end{aligned}$$

Thus the optimal primitive policy μ^* also generates a general policy $\sigma^* = \{\sigma_0^*, \sigma_1^*, \ldots, \sigma_{N-1}^*\}$.

Theorem 2. *Then policy σ^* is optimal in general class; for all $\sigma \in \Pi(g)$*

$$P_{x_0}^{\sigma^*}(G \geq \underline{c}) \quad \geq \quad P_{x_0}^{\sigma}(G \geq \underline{c}) \quad \forall x_0 \in X.$$

Further we have

$$w_0(x_0) \quad = \quad v_0(x_0) \quad \forall x_0 \in X. \tag{12}$$

4. Forward recursive criterion

In this section we introduce the *forward recursive function* :

$$\Psi := \psi_N(\psi_{N-1}(\cdots \psi_1(\psi_0(X_0, U_0); X_1, U_1); \cdots ; X_{N-1}, U_{N-1}); X_N) \tag{13}$$

where

$\psi_0 : X \times U \to R^1$ is an *initial reward function*
$\psi_n : R^1 \times X \times U \to R^1$ is an *n-th forward recursive accumulator*
$\psi_N : R^1 \times X \to R^1$ is a *final reward function.*

Now, let us consider the *forward recursive problem* in $\Pi(g)$:

$$\mathcal{F}_0(x_0) \qquad \begin{array}{ll} \text{Optimize} & P^{\sigma}_{x_0}(\,\Psi \geq \underline{c}) \\ \text{subject to} & (\text{i})_{\text{n}},\ (\text{ii})_{\text{n}}\ \ 0 \leq n \leq N\text{–}1. \end{array}$$

First let us introduce the sequence of one-dimensional *state variables* $\{\lambda_n\}_0^N$ called *forward recursive accumulations* :

$$\begin{aligned} \lambda_0 &:= \underline{\lambda}_0 \ \text{ a fictitious constant} \\ \lambda_n &:= \psi_{n-1}(\cdots(\psi_1(\psi_0(x_0, u_0); x_1, u_1); \cdots ; x_{n-1}, u_{n-1}). \end{aligned}$$

This is equivalent to the sequential dynamics :

$$(\text{i})'_{\text{n}} \quad \lambda_{n+1} = \psi_n(\lambda_n; x_n, u_n) \quad n = 0, \ldots, N\text{–}1, \ \ \lambda_0 = \underline{\lambda}_0.$$

where

$$\psi_0(\lambda_0; x_0, u_0) := \psi_0(x_0, u_0).$$

Then we have the *terminal* expression

$$\psi_N(\psi_{N-1}(\cdots(\psi_1(\psi_0(x_0, u_0); x_1, u_1); \cdots ; x_{N-1}, u_{N-1}); x_N) = \psi_N(\lambda_N; x_N).$$

Second we define the sequence of *forward recursive accumulation sets* $\{\Lambda_n\}$:

$$\Lambda_0 \triangleq \{\lambda_0 \mid \lambda_0 = \underline{\lambda}_0\}$$

$$\Lambda_n \triangleq \left\{ \lambda_n \;\middle|\; \begin{array}{l} \lambda_n = \psi_{n-1}(\cdots(\psi_1(\psi_0(x_0, u_0); x_1, u_1); \cdots ; x_{n-1}, u_{n-1}) \\ (x_0, u_0, \ldots, x_{n-1}, u_{n-1}) \in X \times U \times \cdots \times X \times U \end{array} \right\} \tag{14}$$

$$n = 1, \ldots, N.$$

Thus Λ_n denotes the set of all possible forward recursive accumulations up to n-th stage i.e., for the stage-interval $[0, n)$. Then we have :

Lemma 1. (Forward Recursive Formula)

$$\begin{aligned} \Lambda_0 &= \{\underline{\lambda}_0\} \\ \Lambda_{n+1} &= \{\psi_n(\lambda; x, u) \mid \lambda \in \Lambda_n,\ (x, u) \in X \times U\} \quad 0 \le n \le N\text{–}1. \end{aligned} \tag{15}$$

Proof. Straightforward.

We also define the corresponding *random variables* $\{\widetilde{\Lambda}_n\}_0^N$:

$$\begin{aligned} \widetilde{\Lambda}_0 &:= \underline{\lambda}_0 \\ \widetilde{\Lambda}_n &:= \psi_{n-1}(\cdots(\psi_1(\psi_0(X_0, U_0); X_1, U_1); \cdots; X_{n-1}, U_{n-1}). \end{aligned}$$

This also enjoys the same recursion.

Lemma 2. (Forward Recursive Formula)

$$\begin{aligned} \widetilde{\Lambda}_0 &= \underline{\lambda}_0 \\ \widetilde{\Lambda}_{n+1} &= \psi_n(\widetilde{\Lambda}_n; X_n, U_n) \quad 0 \le n \le N\text{–}1. \end{aligned} \tag{16}$$

Proof. Straightforward.

Finally we construct a new controlled Markov chain on the expanded state spaces $\{X \times \Lambda_n\}_0^N$. Here the state variables $\{(X_n; \widetilde{\Lambda}_n)\}$ behave such that the first component $\{X_n\}$ obeys the original Markov transition law p and the second $\{\widetilde{\Lambda}_n\}$ follows the deterministic dynamics $\lambda_{n+1} := \psi_n(\lambda_n; x_n, u_n)$. When the decision-maker chooses a decision $u_n (\in U)$ on $(x_n; \lambda_n) (\in X \times \Lambda_n)$ at n-th stage, the next state random variable $(X_{n+1}; \widetilde{\Lambda}_{n+1})$ will take $(x_{n+1}; \lambda_{n+1})$ with probability $p(x_{n+1} | x_n, u_n)$ at $(n+1)$-st stage, where $\lambda_{n+1} = \psi_n(\lambda_n; x_n, u_n)$. This is expressed by a coupled dynamics $(\mathrm{i})_\mathrm{n}$, $(\mathrm{i})'_\mathrm{n}$ $0 \le n \le N$–1.

Now we consider the problem of maximizing the *terminal* threshold probability on the expanded Markov chain :

$$\mathcal{F}_0(x_0; \lambda_0) \quad \begin{array}{ll} \text{Maximize} & P^{\gamma}_{x_0, \lambda_0}(\psi_N(\widetilde{\Lambda}_N;\, X_N) \ge \underline{c}) \\ \text{subject to} & (\mathrm{i})_\mathrm{n} \quad X_{n+1} \sim p(\cdot\, | x_n, u_n) \\ & (\mathrm{i})'_\mathrm{n} \quad \lambda_{n+1} = \psi_n(\lambda_n; x_n, u_n) \;\; 0 \le n \le N\text{–}1 \\ & (\mathrm{ii})_\mathrm{n} \quad u_n \in U \end{array}$$

where $\lambda_0 = \underline{\lambda}_0$. Here we take *Markov decision functions* on the expanded state spaces

$$\gamma_n : X \times \Lambda_n \to U.$$

The sequence $\gamma = \{\gamma_0, \ldots, \gamma_{N-1}\}$ is called an *expanded Markov policy* based upon *forward recursive accumulation*. We denote the set of all expanded

Markov policies by $\widetilde{\Pi}$, *expanded Markov class*. For any expanded Markov policy $\gamma(\in \widetilde{\Pi})$, the threshold probability in $\mathcal{F}_0(x_0, \underline{\lambda}_0)$ is the partial multiple sum :

$$P^{\gamma}_{x_0,\lambda_0}(\,\psi_N(\widetilde{\Lambda}_N;\, X_N\,) \geq \underline{c}) \tag{17}$$
$$= \sum\sum\cdots\sum_{(x_1,x_2,\ldots,x_N)\in\star} p(x_1|x_0,u_0)p(x_2|x_1,u_1)\cdots p(x_N|x_{N-1},u_{N-1})$$

where $\star$ is the set

$$\star := \{(x_1,x_2,\ldots,x_N) \mid \psi_N(\lambda_N;\, x_N\,) \geq \underline{c}\} \subset X^N$$

and the sequence of decisions in (17) is determined through γ :

$$u_0 = \gamma_0(x_0;\underline{\lambda}_0),\ u_1 = \gamma_1(x_1;\lambda_1),\ \ldots,\ u_{N-1} = \gamma_{N-1}(x_{N-1};\lambda_{N-1}). \tag{18}$$

We imbed $\mathcal{F}_0(x_0;\lambda_0)$ into the family of subproblems

$$\mathcal{F}_n(x_n;\lambda_n) \qquad \begin{array}{ll} \text{Optimize} & P^{\gamma}_{x_n,\lambda_n}(\,\psi_N(\widetilde{\Lambda}_N;\, X_N\,) \geq \underline{c}) \\ \text{subject to} & (\mathrm{i})_{\mathrm{m}},\ (\mathrm{i})'_{\mathrm{m}},\ (\mathrm{ii})_{\mathrm{m}}\ \ n \leq m \leq N\text{–}1. \end{array}$$

where $(x_n;\lambda_n) \in X{\times}\Lambda_n,\ 0 \leq n \leq N\text{–}1$. Here $P^{\gamma}_{x_n,\lambda_n}(\cdot)$ is the probability measure on expanded state spaces $\{X{\times}\Lambda_m\}_n^N$ which is defined uniquely through *initial state* (x_n,λ_n), expanded Markov policy $\gamma = \{\gamma_n,\ \ldots,\ \gamma_{N-1}\}$ and the Markov transition law from n-th stage on.

Let $w^n(x_n;\lambda_n)$ be the maximum value, where

$$w^N(x_N;\lambda_N) := 1_{\underline{c}}(\psi_N(\lambda_N;x_N)) \quad (x_N;\lambda_N) \in X{\times}\Lambda_N.$$

Then we have :

Theorem 3. (Backward Recursive Equation)

$$w^n(x;\lambda) = \mathop{\mathrm{Max}}_{u\in U} \sum_{y\in X} w^{n+1}(y;\psi_n(\lambda;x,u))p(y|x,u) \tag{19}$$
$$(x;\lambda) \in X{\times}\Lambda_n,\ \ 0 \leq n \leq N\text{–}1$$

$$w^N(x;\lambda) = 1_{\underline{c}}(\psi_N(\lambda;x)) \quad (x;\lambda) \in X{\times}\Lambda_N. \tag{20}$$

Proof. It is straightforward.

Let $\gamma_n^*(x;\lambda)$ denote a maximizer in (19). Then we have an optimal policy $\gamma^* = \{\gamma_0^*,\cdots,\gamma_{N-1}^*\}$ in $\widetilde{\Pi}$. Further γ^* generates a general policy $\sigma^* = \{\sigma_0^*,\ldots,\sigma_{N-1}^*\}$, where $\sigma_n^*(x_0,x_1,\ldots,x_n)$ is specified as follows :

$$
\begin{aligned}
&u_0 := \gamma_0^*(x_0;\underline{\lambda}_0), \;\; \lambda_1 := \psi_0(\underline{\lambda}_0;x_0,u_0) \\
&u_1 := \gamma_1^*(x_1;\lambda_1), \;\; \lambda_2 := \psi_1(\lambda_1;x_1,u_1) \\
&\qquad\qquad\vdots \\
&u_{n-1} := \gamma_{n-1}^*(x_{n-1};\lambda_{n-1}), \;\; \lambda_n := \psi_{n-1}(\lambda_{n-1};x_{n-1},u_{n-1}) \\
&\sigma_n^*(x_0,x_1,\ldots,x_n) := \gamma_n^*(x_n;\lambda_n).
\end{aligned} \tag{21}
$$

Then we have :

Theorem 4.
(i) *The policy σ^* is optimal in general class $\Pi(g)$.*
(ii) *The maximum value in expanded Markov class $\widetilde{\Pi}$ is equal to the maximum value in general class $\Pi(g)$:*

$$w^0(x_0;\underline{\lambda}_0) = v_0(x_0). \tag{22}$$

Proof. Using the same discussion as in §5 Equivalences and Optimality of [8], we may complete the proof of Theorem 4. The crucial point in the above discussion is that the maximum attained by a general policy can also be attained by an expanded Markov policy. This is shown by use of primitive policy as a bypass. Then an essential assumption is the forward recursiveness; the past is separable in a reward function. A full statement takes the same space as in [8, §5].

5. Backward recursive criterion

Let us now define the *backward recursive function* :

$$\Phi := \phi_0(X_0,U_0;\phi_1(X_1,U_1;\cdots;\phi_{N-1}(X_{N-1},U_{N-1};k(X_N))\cdots)). \tag{23}$$

where

$\phi_n : X{\times}U{\times}R^1 \to R^1$ is an *n-th backward recursive accumulator*
$k : X \to R^1$ is a *terminal reward function.*

We consider the *backward recursive problem* in $\Pi(g)$:

$$
\mathcal{B}_0(x_0) \quad
\begin{array}{lll}
\text{Maximize} & P_{x_0}^{\sigma}(\,\Phi \geq \underline{c}\,) & \\
\text{subject to} & \text{(i)}_\text{n} \;\; X_{n+1} \sim p(\,\cdot\,|x_n,u_n) & \\
& & n = 0,\ldots,N{-}1 \\
& \text{(ii)}_\text{n} \;\; u_n \in U &
\end{array}
$$

We assume that each $\phi_n(x,u;\cdot)$ is continuous strictly increasing function of its third argument. So it is invertible. The invese function is written as

$$\phi_n^{-1}(x,u)(\cdot) \text{ or } \phi_n^{-1}(x,u;\cdot).$$

First let us introduce the sequence of additional one-dimensional state variables $\{c_n\}_0^N$ called *threshold levels* :

$$\begin{aligned} c_0 &:= \underline{c} \\ c_n &:= \phi_{n-1}^{-1}(x_{n-1},u_{n-1}) \circ \cdots \circ \phi_1^{-1}(x_1,u_1) \circ \phi_0^{-1}(x_0,u_0)(\underline{c}). \end{aligned}$$

This is equivalent to the sequential dynamics :

$$(\mathrm{i})_n'' \quad c_{n+1} = \phi_n^{-1}(x_n,u_n;c_n) \quad n = 0,\ldots,N{-}1, \quad c_0 = \underline{c}.$$

It turns out that under the sequential constraint

$$\phi_0(x_0,u_0;\phi_1(x_1,u_1;\cdots;\phi_{N-1}(x_{N-1},u_{N-1};k(x_N))\cdots)) \geq \underline{c}$$

if and only if

$$k(x_N) \geq c_N.$$

This would yield the same probability

$$P(\Phi \geq \underline{c}) \;=\; P(\,k(X_N) \geq c_N).$$

Thus the introduction of threshold levels has reduced the *original* threshold probability in $\mathcal{B}_0(x_0)$ to the same *terminal* one.

Second we define *threshold level sets* $\{C_n\}$:

$$\begin{aligned} C_0 &\triangleq \{c_0 \mid c_0 = \underline{c}\} \quad \text{where } \underline{c} \text{ is the given lower level} \\ C_n &\triangleq \left\{ c_n \;\middle|\; \begin{array}{c} c_n = \phi_{n-1}^{-1}(x_{n-1},u_{n-1}) \circ \cdots \circ \phi_0^{-1}(x_0,u_0)(\underline{c}) \\ (x_0,u_0,\ldots,x_{n-1},u_{n-1}) \in X{\times}U{\times}\cdots{\times}X{\times}U \end{array} \right\} \qquad (24) \\ &\qquad\qquad n = 1,\ldots,N. \end{aligned}$$

Thus C_n denotes the set of all possible threshold levels for the future process on stage-interval $[n,N]$. Then we have the forward recursive formula :

Lemma 3.

$$\begin{aligned} C_0 &= \{\underline{c}\} \\ C_{n+1} &= \{\phi_n^{-1}(x,u;c) \mid c \in C_n,\ (x,u) \in X{\times}U\} \quad 0 \leq n \leq N-1. \end{aligned} \qquad (25)$$

Finally we introduce a new controlled Markov chain on the expanded state spaces $\{X{\times}C_n\}_0^N$. Here the state variables $\{(X_n;c_n)\}$ behave such that the first component $\{X_n\}$ obeys the original Markov transition law p and the second follows the deterministic dynamics $c_{n+1} = \phi_n^{-1}(x_n,u_n;c_n)$. When the decision-maker chooses a decision $u_n(\in U)$ on $(x_n;c_n)(\in X{\times}C_n)$ at n-th stage, the next state random variable $(X_{n+1}; c_{n+1})$ will take $(x_{n+1}; c_{n+1})$ with probability $p(x_{n+1}|x_n, u_n)$ at $(n+1)$-st stage, where $c_{n+1} = \phi_n^{-1}(x_n, u_n; c_n)$. Thus this is the coupled dynamics $(\mathrm{i})_\mathrm{n}$, $(\mathrm{i})''_\mathrm{n}$ $0 \le n \le N-1$.

Now we maximize the *terminal* threshold probability on the expanded Markov chain :

$$\mathcal{B}_0(x_0, \underline{c}) \quad \begin{array}{ll} \text{Maximize} & P^{\tau}_{x_0,\underline{c}}(\, k(X_N) \ge c_N) \\ \text{subject to} & (\mathrm{i})_\mathrm{n} \quad X_{n+1} \sim p(\,\cdot\,|x_n, u_n) \\ & (\mathrm{i})''_\mathrm{n} \quad c_{n+1} = \phi_n^{-1}(x_n, u_n; c_n), \\ & \qquad\qquad\qquad c_0 = \underline{c} \quad n = 0, \ldots, N\text{–}1 \\ & (\mathrm{ii})_\mathrm{n} \quad u_n \in U \end{array}$$

Let $\tau = \{\tau_0, \ldots, \tau_{N-1}\}$ be a sequence of Markov decision functions

$$\tau_n : X \times C_n \to U$$

on the expanded state spaces. Then τ is called an *expanded Markov policy* based upon *threshold levels*. The set of all expanded Markov policies is denoted by $\widehat{\Pi}$. For any $\tau (\in \widehat{\Pi})$, threshold probability is the partial multiple sum :

$$P^{\tau}_{x_0,\underline{c}}(\, k(X_N) \ge c_N\,) = \sum\sum\cdots\sum_{(x_1,\ldots,x_N)\,;\,k(x_N)\ge c_N} p(x_1|x_0, u_0)p(x_2|x_1, u_1)\cdots p(x_N|x_{N-1}, u_{N-1}) \tag{26}$$

where the decisions in (26) is determined through τ :

$$u_0 = \tau_0(x_0; c_0),\ u_1 = \tau_1(x_1; c_1),\ \ldots,\ u_{N-1} = \tau_{N-1}(x_{N-1}; c_{N-1}). \tag{27}$$

Now we consider the subprocess from n-th stage to N-th governed by an expanded Markov policy $\tau = \{\tau_n, \ldots, \tau_{N-1}\}$. $\widehat{\Pi}(n)$ denotes the set of all such policies. Hence, $\widehat{\Pi}(0) = \widehat{\Pi}$. We note that the terminal probability of $k(X_N) \ge c_N$ under the condition $X_N = x_N$ becomes

$$P(k(X_N) \ge c_N \,|\, X_N = x_N) = 1_{c_N}(k(x_N)) \quad x_N \in X. \tag{28}$$

Lemma 4. *We have for any* $0 \le n \le N\text{–}1$, $(x_n; c_n) \in X \times C_n$ *and* $\tau = \{\tau_n, \ldots, \tau_{N-1}\} \in \widehat{\Pi}(n)$

$$P^{\tau}_{x_n,c_n}(\, k(X_N) \ge c_N) \quad = \quad \sum_{x_{n+1}\in X} P^{\tau'}_{x_{n+1},c_{n+1}}(\, k(X_N) \ge c_N)p(x_{n+1}|x_n, u_n)$$

where

$$c_{n+1} = \phi_n^{-1}(x_n, u_n; c_n), \quad u_n = \tau_n(x_n; c_n), \quad \tau' = \{\tau_{n+1}, \ldots, \tau_{N-1}\},$$

and $P^{\tau'}_{x_N,c_N} := P$ *in (28) for* $\tau = \{\tau_{N-1}\}$.

Proof. It suffices to verify the equality

$$\sum\sum\cdots\sum_{(x_{n+1},\ldots,x_N)\,;\,k(x_N)\geq c_N} p_{n+1}p_{n+2}\cdots p_N$$

$$= \sum_{x_{n+1}\in X}\left[\sum\sum\cdots\sum_{(x_{n+2},\ldots,x_N)\,;\,k(x_N)\geq c_N} p_{n+2}\cdots p_N\right] p_{n+1},$$

where

$$p_{m+1} = p(x_{m+1}|x_m, u_m),\ u_m = \tau_m(x_m; c_m),\ c_{m+1} = \phi_m^{-1}(x_m, u_m; c_m).$$

Now let us imbed $\mathcal{B}_0(x_0; \underline{c})$ into the family of subproblems $\{\mathcal{B}_n(x_n; c_n)\}$, where $\mathcal{B}_n(x_n; c_n)$ is the controlled Markov chain starting at state $(x_n; c_n)$ from n-th stage on :

$$\mathcal{B}_n(x_n; c_n) \qquad \begin{array}{ll} \text{Maximize} & P^{\tau}_{x_n,c_n}(\,k(X_N) \geq c_N) \\ \text{subject to} & (\mathrm{i})_{\mathrm{m}},\ (\mathrm{i})''_{\mathrm{m}},\ (\mathrm{ii})_{\mathrm{m}}\ \ m = n, \ldots, N\text{–}1. \end{array}$$

Here the maximization is taken over $\widehat{\Pi}(n)$. We see that the restoration of $\{(\mathrm{i})''_{\mathrm{m}}\}$ to reward accumulation yields an equivalent nonterminal (forward recursive) thershold probability form :

$$\begin{array}{ll} \text{Maximize} & P^{\tau}_{x_n}(\,\Phi_n \geq c_n) \\ \text{subject to} & (\mathrm{i})_{\mathrm{m}},\ (\mathrm{ii})_{\mathrm{m}} \qquad m = n, \ldots, N\text{–}1. \end{array}$$

where

$$\Phi_n := \phi_n(X_n, U_n; \cdots; \phi_{N-1}(X_{N-1}, U_{N-1}; k(X_N))\cdots).$$

Let $f_n(x_n; c_n)$ denote the maximum value of $\mathcal{B}_n(x_n; c_n)$, where

$$f_N(x_N; c_N) := P(k(X_N) \geq c_N \,|\, X_N = x_N).$$

Then we have the backward recursive equation :

Theorem 5.

$$f_n(x; c) = \operatorname*{Max}_{u\in U} \sum_{y\in X} f_{n+1}(y; \phi_n^{-1}(x, u; c))p(y|x, u) \tag{29}$$

$$(x; c) \in X{\times}C_n,\ \ 0 \leq n \leq N\text{–}1$$

$$f_N(x; c) \;=\; 1_c(k(x)) \qquad (x; c) \in X{\times}C_N. \tag{30}$$

Let $\bar{\tau}_n(x;c)$ denote a maximizer in (29). Then we have an optimal policy $\bar{\tau} = \{\bar{\tau}_0, \ldots, \bar{\tau}_{N-1}\}$ in expanded Markov class $\widehat{\Pi}$. Further $\bar{\tau}$ generates a general policy $\bar{\sigma} = \{\bar{\sigma}_0, \ldots, \bar{\sigma}_{N-1}\}$, which defines $\bar{\sigma}_n(x_0, x_1, \ldots, x_n)$ as follows :

$$\begin{aligned} u_0 &:= \bar{\tau}_0(x_0; \underline{c}), \quad c_1 := \phi_0^{-1}(x_0, u_0; \underline{c}) \\ u_1 &:= \bar{\tau}_1(x_1; c_1), \quad c_2 := \phi_1^{-1}(x_1, u_1; c_1) \\ &\vdots \\ u_{n-1} &:= \bar{\tau}_{n-1}(x_{n-1}; c_{n-1}), \quad c_n := \phi_{n-1}^{-1}(x_{n-1}, u_{n-1}; c_{n-1}) \\ \bar{\sigma}_n(x_0, x_1, \ldots, x_n) &:= \bar{\tau}_n(x_n; c_n). \end{aligned} \tag{31}$$

Then we have the following result:

Theorem 6.
(i) *The policy* $\bar{\sigma}$ *is optimal in general class* $\Pi(g)$.
(ii) *The maximum value of expanded Markov class* $\widehat{\Pi}$ *is equal to the maximum value of general class* $\Pi(g)$:

$$f_0(x_0; \underline{c}) = v_0(x_0). \tag{32}$$

Proof. This is also shown by the similar fashion as for Theorem 4 (see also [8, Theorem 5.2]).

Concluding Remark. As a probability criterion, we have chosen the threshold probability with a lower level $\underline{c}$. Needless to say, there are many types of probability criterion. A frequent usage is the sandwich probability, which is characterized by both the lower level $\underline{c}$ and an upper level $\overline{c}$. The sandwich probability $P^{\sigma}_{x_0}(\,\underline{c} \leq G < \overline{c}\,)$ is optimized by adding two state variables (that is, a lower level and an upper level) for the backward recursive criterion. In general, for any given subset $E(\subset R^1)$, the probability $P^{\sigma}_{x_0}(\,G \in E\,)$ is also optimized both general and forward recursive criteria. However, the backward optimization may require an infinitely many new state variables.

Acknowledgements. The author would like to thank an anonymous referee for his useful advices and careful reading of this paper.

References

1. Bellman, R.E.: Dynamic Programming. Princeton University Press, Princeton, NJ 1957

2. Bellman, R.E.: Some Vistas of Modern Mathematics. University of Kentucky Press, Lexington, KY 1968
3. Blackwell, D.: Discounted dynamic programming. Ann. Math. Stat. **36**, 226-235 (1965)
4. Denardo, E.V.: Contraction mappings in the theory underlying dynamic programming. SIAM Review **9**, 165-177 (1968)
5. Iwamoto, S.: Conditional decision processes with recursive reward function. J. Math. Anal. Appl. **230**, 193-210 (1999)
6. Iwamoto, S.: Recursive method in stochastic optimization under compound criteria. Adv. Math. Econ. **3**, 63-82 (2001)
7. Iwamoto, S., Tsurusaki, K., Fujita, T.: Conditional decision-making in a fuzzy environment. J. Operations Res. Soc. Japan **42**, 198-218 (1999)
8. Iwamoto, S., Ueno, T., Fujita, T.: Controlled Markov chains with utility functions. In: Markov Processes and Controlled Markov Chains (Z.Hou et al. eds.), pp.135-148 Kluwer, Dordrecht 2002
9. Le Van, C., Dana, D.-A.: Dynamic Programming in Economics. Kluwer, Boston, MA 1993
10. Ozaki, H., Streufert, P.A.: Dynamic programming for non-additive stochastic objects. J. Math. Eco. **25**, 391-442 (1996)
11. Sargent, T.J.: Dynamic Macroeconomic Theory. Harvard University Press, Cambridge, MA 1987
12. Stokey, N.L., Lucas, R.E.: Recursive Methods in Economic Dynamics. Harvard University Press, Cambridge, MA 1989
13. Strauch, R.: Negative dynamic programming. Ann. Math. Stat. **37**, 871-890 (1966)
14. Streufert, P.A.: Recursive Utility and Dynamic Programming. In: Handbook of Utility Theory Vol. 1 (S. Barberà et al. eds.), Chap. III. Kluwer, Boston, MA 1987

Adv. Math. Econ. 6, 69–83 (2004)

Advances in
MATHEMATICAL
ECONOMICS

Approximation of expectation of diffusion processes based on Lie algebra and Malliavin calculus

Shigeo Kusuoka

Graduate School of Mathematical Sciences, The University of Tokyo, 3-8-1 Komaba, Meguro-ku, Tokyo 153-8914, Japan

Received: July 9, 2003
Revised: September 17, 2003

JEL classification: C63, G12

Mathematics Subject Classification (2000): 65C05, 60G40

Abstract. The author gives a new numerical computation method of Expectation of Diffusion Processes, which is an improvement of a results in [3].

Key words: mathematical finance, option pricing, Lie algebra

1. Introduction

It is important to compute expectations of diffusion processes numerically, in the case when we apply mathematical finance to practical problems. There are a lot of works in this field (cf.. Ballay and Talay [1], Kloeden and Platen [2]). The author gave a new method in [3] and some related works have already appeared (Lyons and Victoir [7], Ninomiya [8]).

In the present paper, we refine and extend the idea in [3] by using notions in [10]. We use the notation in [10] for free Lie algebra. Let $(\Omega, \mathcal{F}, P)$ be a probability space and let $\{(B^1(t), \ldots, B^d(t); t \in [0, \infty)\}$ be a d-dimensional Brownian motion. Let $B^0(t) = t$, $t \in [0, \infty)$. Let $V_0, V_1, \ldots, V_d \in C_b^\infty(\mathbf{R}^N; \mathbf{R}^N)$. Here $C_b^\infty(\mathbf{R}^N; \mathbf{R}^n)$ denotes the space of $\mathbf{R}^n$-valued smooth functions defined in $\mathbf{R}^N$ whose devivatives of any order are bounded. We regard elements in $C_b^\infty(\mathbf{R}^N; \mathbf{R}^N)$ as vector fields on $\mathbf{R}^N$.

Now let $X(t, x)$, $t \in [0, \infty)$, $x \in \mathbf{R}^N$, be the solution to the Stratonovich stochastic integral equation

$$X(t,x) = x + \sum_{i=0}^{d} \int_0^t V_i(X(s,x)) \circ dB^i(s). \tag{1}$$

Then there is a unique solution to this equation. Moreover we may assume that with probability one $X(t,x)$ is continuous in t and smooth in x.

Let $A = A_d = \{v_0, v_1, \ldots, v_d\}$, be an alphabet, a set of letters, and A^* be the set of words consisting of A including the empty word which is denoted by 1. For $u = u^1 \cdots u^k \in A^*$, $u^j \in A$, $j = 1, \ldots, k$, $k \geq 0$, we denote by $n_i(u)$, $i = 0, \ldots, d$, the cardinal of $\{j \in \{1, \ldots, k\}; u^j = v_i\}$. Let $|u| = n_0(u) + \ldots + n_d(u)$, a length of u, and $\| u \| = |u| + n_0(u)$ for $u \in A^*$. Let $\mathbf{R}\langle A \rangle$ be the $\mathbf{R}$-algebra of noncommutative polynomials on A, $\mathbf{R}\langle\langle A \rangle\rangle$ be the $\mathbf{R}$-algebra of noncommutative formal series on A, $\mathcal{L}(A)$ be the free Lie algebra over $\mathbf{R}$ on the set A, and $\mathcal{L}((A))$ be the $\mathbf{R}$ Lie algebra of free Lie series on the set A.

Let ι denotes the left normed bracketing operator, i.e.,

$$\iota(v_{i_1} \cdots v_{i_n}) = [\ldots [v_{i_1}, v_{i_2}], \ldots, v_{i_n}].$$

For any $w_i = \sum_{u \in A^*} a_{iu} u, \in \mathbf{R}\langle A \rangle$, $i = 1, 2$, let us define an inner product $\langle w_1, w_2 \rangle$ and a norm $\| w_1 \|_2$ by

$$\langle w_1, w_2 \rangle = \sum_{u \in A^*} a_{1u} a_{2u} \in \mathbf{R} \text{ and } \| w_1 \|_2 = (\langle w_1, w_1 \rangle)^{1/2}.$$

We can regard vector fields $V_0, V_1, \ldots, V_d$ as first differential operators over $\mathbf{R}^N$. Let $\mathcal{DO}(\mathbf{R}^N)$ denotes the set of smooth differential operators over $\mathbf{R}^N$. Then $\mathcal{DO}(\mathbf{R}^N)$ is a noncommutative algebra over $\mathbf{R}$. Let $\Phi : \mathbf{R}\langle A \rangle \to \mathcal{DO}(\mathbf{R}^N)$ be a homomorphism given by

$$\Phi(1) = Identity, \qquad \Phi(v_{i_1} \cdots v_{i_n}) = V_{i_1} \cdots V_{i_n}, \quad n \geq 1,\ i_1, \ldots, i_n = 0, 1, \ldots, d.$$

Then we see that

$$\Phi(\iota(v_{i_1} \cdots v_{i_n})) = [\cdots [V_{i_1}, V_{i_2}], \cdots, V_{i_n}], \quad n \geq 2,\ i_1, \ldots, i_n = 0, 1, \ldots, d.$$

Let $B(t; u)$, $t \in [0, \infty)$, $u \in A^*$, be inductively defined by

$$B(t; 1) = 1, \qquad B(t; v_i) = B^i(t),\ i = 0, 1, \ldots, d,$$

and

$$B(t; uv_i) = \int_0^t B(s; u) \circ dB^i(s) \qquad u \in A^*,\ i = 0, \ldots, d.$$

Also we define $B(t;w)$ $t \in [0,\infty)$, $w \in \mathbf{R}\langle A\rangle$ by

$$B(t; \sum_{u\in A^*} a_u u) = \sum_{u\in A^*} a_u B(t;u).$$

Let $A_m^* = \{u \in A^*;\ \| u \| = m\}$, $m \geq 0$, and let $\mathbf{R}\langle A\rangle_m = \sum_{u\in A_m^*} \mathbf{R}u$, and $\mathbf{R}\langle A\rangle_{\leq m} = \sum_{k=0}^m \mathbf{R}\langle A\rangle_k$, $m \geq 0$. Let $j_m : \mathbf{R}\langle\langle A\rangle\rangle \to \mathbf{R}\langle A\rangle_{\leq m}$ be a natural sujective linear map such that $j_m(u) = u$, $u \in A^*$, $\| u \| \leq m$, and $j_m(u) = 0$, $u \in A^*$, $\| u \| \geq m+1$. Let $\mathcal{L}(A)_m = \mathcal{L}(A) \cap \mathbf{R}\langle A\rangle_m$, and $\mathcal{L}(A)_{\leq m} = \mathcal{L}(A) \cap \mathbf{R}\langle A\rangle_{\leq m}$, $m \geq 1$. Let $A^{**} = \{u \in A^*;\ u \neq 1, v_0\}$, and $A^{**}_{\leq m} = \{u \in A^{**};\ \| u \| \leq m\}$, $m \geq 1$.

Let $\Psi_s : \mathbf{R}\langle\langle A\rangle\rangle \to \mathbf{R}\langle\langle A\rangle\rangle$, $s > 0$, be given by

$$\Psi_s(\sum_{m=0}^{\infty} x_m) = \sum_{m=0}^{\infty} s^{m/2} x_m, \qquad x_m \in \mathbf{R}\langle A\rangle_m, \quad m \geq 0.$$

Now we introduce a condition (UFG) on the family of vector field $\{V_0, V_1, \ldots, V_d\}$ as follows.
(UFG) There are an integer ℓ and $\varphi_{u,u'} \in C_b^\infty(\mathbf{R}^N)$, $u \in A^{**}$, $u' \in A^{**}_{\leq \ell}$, satisfying the following.

$$\Phi(\iota(u)) = \sum_{u'\in A^{**}_{\leq \ell}} \varphi_{u,u'} \Phi(\iota(u')), \qquad u \in A^{**}.$$

Let us define a semi-norm $\| \cdot \|_{V,n}$, $n \geq 1$, on $C_b^\infty(\mathbf{R}^N;\mathbf{R})$ by

$$\| f \|_{V,n} = \sum_{k=1}^{n} \sum_{u_1,\ldots,u_k\in A^{**}, \|u_1\cdots u_k\|=n} \| \Phi(\iota(u_1)\cdots\iota(u_k))f \|_\infty .$$

Here $\| f \|_\infty = \sup\{|f(x)|;\ x \in \mathbf{R}^N\}$.

Now let us define a semigroup of linear operators $\{P_t\}_{t\in[0,\infty)}$ by

$$(P_t f)(x) = E[f(X(t,x))], \qquad t \in [0,\infty),\ f \in C_b^\infty(\mathbf{R}^N).$$

Let us think of a family $\{Q_{(s)};\ s \in (0,1]\}$ of linear operators in $C_b(\mathbf{R}^N)$.

Definition 1. *We say that $Q_{(s)}$, $s \in (0,1]$, is m-similar, $m \geq 1$, if there are a constant $C > 0$ and $M \geq m+1$ such that*

$$\| P_s f - Q_{(s)} f \|_\infty \leq C(\sum_{k=m+1}^{M} s^{k/2} \| f \|_{V,k} + s^{(m+1)/2} \| \nabla f \|_\infty),$$

$$\| Q_{(s)} f - P_s f \|_\infty \leq C s^{1/2} \| \nabla f \|_\infty,$$

and

$$\| Q_{(s)} f \|_\infty \leq \exp(Cs) \| f \|_\infty$$

for any $s \in (0,1]$, and $f \in C_b^\infty(\mathbf{R}^N;\mathbf{R})$.

Definition 2. (1) *We say that an $\mathcal{L}((A))$-valued random variable ξ is $L^{\infty-}$, if*

$$E[\| j_n(\xi) \|_2^n] < \infty \qquad \textit{for any} \quad n \geq 1.$$

(2) *We say that an $\mathcal{L}((A))$-valued random variable ξ is m-$\mathcal{L}$-moment similar, $m \geq 2$, if $j_m(\xi)$ is $L^{\infty-}$,*

$$\langle \xi, v_0 \rangle = 1 \quad a.s.,$$

and if

$$E[j_m(\exp(\xi))] = E[j_m(X(1))].$$

Here $X(t)$ is a solution to the SDE (2) *on $\mathbf{R}\langle\langle A \rangle\rangle$ in Section 3.*

Our main results are the following.

Theorem 3. *Let $m \geq 1$ and ξ be an $\mathcal{L}((A))$-valued m-$\mathcal{L}$-moment similar random variable. Also, let $Y : (0,1] \times \mathbf{R}^N \times \Omega \to \mathbf{R}^N$ be a measurable map such that $Y(s, \cdot, \omega) : \mathbf{R}^N \to \mathbf{R}^N$ is continuous for any $s \in (0,1]$ and $\omega \in \Omega$, and*

$$\sup_{s \in (0,1], x \in \mathbf{R}^N} s^{-(m+1)/2} E[|Y(s,x)|] < \infty.$$

Let us define linear operators $Q_{(s)}$, $s > 0$, in $C_b(\mathbf{R}^N)$ by

$$(Q_{(s)} f)(x) = E[f(\exp(\Phi(j_m(\Psi_s(\xi))))(x) + Y(s,x))], \qquad f \in C_b(\mathbf{R}^N).$$

Then $\{Q_{(s)};\ s \in (0,1]\}$ is m-similar.

Theorem 4. *Assume that the family of vector fields satisfies the condition* (UFG). *Let $m \geq 1$ and $Q_{(s)}$, $s > 0$, be an m-similar family of linear operators in $C_b(\mathbf{R}^N)$. Also, let $T > 0$ and $\gamma > 0$, $t_k = t_k^{(n)} = \dfrac{k^\gamma T}{n^\gamma}$, $n \geq 1$, $k = 0, 1, \ldots, n$, and let $s_k = s_k^{(n)} = t_k - t_{k-1}$, $k = 1, \ldots, n$. Then we have the following.*
For $\gamma \in (0, m-1)$, there is a constant $C > 0$ such that

$$\| P_T f - Q_{(s_n)} Q_{(s_{n-1})} \cdots Q_{(s_1)} f \|_\infty \leq C n^{-\gamma/2} \| \nabla f \|_\infty, \qquad f \in C_b^\infty(\mathbf{R}^N),\ n \geq 1.$$

For $\gamma = m-1$, there is a constant $C > 0$ such that

$$\| P_T f - Q_{(s_n)} Q_{(s_{n-1})} \cdots Q_{(s_1)} f \|_\infty \leq C n^{-\frac{m-1}{2}} \log(n+1) \| \nabla f \|_\infty, \qquad f \in C_b^\infty(\mathbf{R}^N),\ n \geq 1.$$

For $\gamma > m-1$, there is a constant $C > 0$ such that

$$\| P_T f - Q_{(s_n)} Q_{(s_{n-1})} \cdots Q_{(s_1)} f \|_\infty \leq C n^{-\frac{m-1}{2}} \| \nabla f \|_\infty, \qquad f \in C_b^\infty(\mathbf{R}^N),\ n \geq 1.$$

2. Proof of Theorem 4

First, note the following (cf. [4]).

Theorem 5. *Assume that the family of vector fields satisfies the condition* (UFG). *Then for any $n \geq 2$ there is a constant $C > 0$ such that*

$$\| P_t f \|_{V,n} \leq \frac{C}{t^{(n-1)/2}} \| \nabla f \|_\infty, \quad f \in C_b^\infty(\mathbf{R}^N),\ t \in (0,1].$$

Now let us prove Theorem 4. Note that for $k = 2, \ldots, n$, and $\ell \geq m+1$,

$$\begin{aligned}\frac{s_k^{\ell/2}}{t_{k-1}^{(\ell-1)/2}} &= T^{1/2} \frac{(\int_{k-1}^k \gamma s^{\gamma-1} ds)^{\ell/2}}{n^{\gamma/2}(k-1)^{(\ell-1)\gamma/2}} \\ &\leq T^{1/2}\gamma^\ell n^{-\gamma/2}(k-1)^{(\gamma-\ell)/2}((\frac{k}{k-1})^{\gamma-1} \vee 1).\end{aligned}$$

So we have

$$\begin{aligned}&\| P_T f - Q_{(s_n)} \cdots Q_{(s_1)} f \|_\infty \\ &\leq \sum_{k=1}^n \| Q_{(s_n)} \cdots Q_{(s_{k+1})} P_{t_k} f - Q_{(s_n)} \cdots Q_{(s_k)} P_{t_{k-1}} f \|_\infty \\ &\leq e^{CT} \sum_{k=1}^n \| P_{s_k} P_{t_{k-1}} f - Q_{(s_k)} P_{t_{k-1}} f \|_\infty \\ &\leq Ce^{CT} (\sum_{k=2}^n (\sum_{\ell=m+1}^M s_k^{\ell/2} \| P_{t_{k-1}} f \|_{V,\ell} + s_k^{(m+1)/2} \| \nabla P_{t_{k-1}} f \|_\infty) \\ &\qquad\qquad + s_1^{(m+1)/2} \| \nabla f \|_\infty) \\ &\leq C_1 (\sum_{k=2}^n (\sum_{\ell=m+1}^M \frac{s_k^{\ell/2}}{t_{k-1}^{(\ell-1)/2}}) + \sum_{k=1}^n s_k^{(m+1)/2}) \| \nabla f \|_\infty \\ &\leq C_2 (n^{-\gamma/2} \sum_{k=2}^n (k-1)^{(\gamma-(m+1))/2} + n^{-(m+1)/2}) \| \nabla f \|_\infty .\end{aligned}$$

So we have the assertions in Theorem 4.

3. Algebraic structure of iterated integrals

We define a metric function dis over $\mathbf{R}\langle\langle A \rangle\rangle$ by

$$dis(w_1, w_2) = \sum_{u \in A^*} (d+2)^{-|u|}(1 \wedge |a_{1,u} - a_{2,u}|)$$

for $w_i = \sum_{u \in A^*} a_{i,u}u$, $i = 1, 2$, $a_{i,u} \in \mathbf{R}$, $u \in A^*$. Then $\mathbf{R}\langle\langle A \rangle\rangle$ becomes a Polish space. Let $\mathcal{B}(\mathbf{R}\langle\langle A \rangle\rangle)$ be a Borel algebra over $\mathbf{R}\langle\langle A \rangle\rangle$.

Let $(\Omega, \mathcal{F}, P)$ be a complete probability space. One can define $\mathbf{R}\langle\langle A \rangle\rangle$-valued random variables and their expectaions etc. naturally. Let $\{\mathcal{F}_t\}_{t \in [0,\infty)}$ be a filtration satisfying a usual hypothesis, $(B^1(t), \dots, B^d(t))$, $t \in [0, \infty)$, be a d-dimensional $\{\mathcal{F}_t\}_{t \in [0,\infty)}$-Brownian motion, and $B^0(t) = t$, $t \in [0, \infty)$. We say that $X(t)$ is an $\mathbf{R}\langle\langle A \rangle\rangle$-valued continuous semimartingale, if there are continuous semimartingales X_u, $u \in A^*$, such that $X(t) = \sum_{u \in A^*} X_u(t)u$. For $\mathbf{R}\langle\langle A \rangle\rangle$-valued continuous semimartingale $X(t), Y(t)$, we can define $\mathbf{R}\langle\langle A \rangle\rangle$-valued continuous semimartingales $\int_0^t X(s) \circ dY(s)$ and $\int_0^t \circ dX(s)Y(s)$ by

$$\int_0^t X(s) \circ dY(s) = \sum_{u,w \in A^*} (\int_0^t X_u(s) \circ dY_w(s))uw,$$

$$\int_0^t \circ dX(s)Y(s) = \sum_{u,w \in A^*} (\int_0^t Y_w(s) \circ dX_u(s))uw,$$

where

$$X(t) = \sum_{u \in A^*} X_u(t)u, \qquad Y(t) = \sum_{w \in A^*} Y_w(t)w.$$

Then we have

$$X(t)Y(t) = X(0)Y(0) + \int_0^t X(s) \circ dY(s) + \int_0^t \circ dX(s)Y(s).$$

Since $\mathbf{R}$ is regarded a vector subspace in $\mathbf{R}\langle\langle A \rangle\rangle$, we can define $\int_0^t X(s) \circ dB^i(s)$, $i = 0, 1, \dots, d$, naturally.

Now let us consider the following SDE on $\mathbf{R}\langle\langle A \rangle\rangle$

$$X(t) = 1 + \sum_{i=0}^d \int_0^t X(s)v_i \circ dB^i(s), \qquad t \geq 0. \tag{2}$$

One can easily solve this SDE and obtains

$$X(t) = \sum_{u \in A^*} B(t; u)u.$$

We also have the following.

Proposition 6. $\log X(t) \in \mathcal{L}((A))$, $t \geq 0$, *with probability one.*

Proof. Note that

$$\delta(X(t)) = 1 \otimes 1 + \sum_{i=0}^{d} \int_0^t \delta(X(s))(v_i \otimes 1 + 1 \otimes v_i) \circ dB^i(s),$$

and

$$\begin{aligned} X(t) \otimes X(t) = 1 \otimes 1 + \int_0^t \circ d(X(s) \otimes 1)(1 \otimes X(s)) \\ + \int_0^t (X(s) \otimes 1) \circ d(1 \otimes X(s)) \\ = 1 \otimes 1 + \sum_{i=0}^{d} \int_0^t X(s) \otimes X(s)(v_i \otimes 1 + 1 \otimes v_i) \circ dB^i(s). \end{aligned}$$

Here δ is the coproduct (see [10] p.19). Since one can easily see the uniqueness of such SDE on $\mathbf{R}\langle\langle A\rangle\rangle$, we have

$$\delta(X(t)) = X(t) \otimes X(t).$$

Then we have our assertion from [10] Theorem 3.2. □

Proposition 7. *For any* $m, n \geq 1$, *and* $x \in \mathcal{L}((A))$ *with* $\langle x, 1\rangle = 0$,

$$j_m(\pi_n \exp(x)) = \pi_n(j_m \exp(x)).$$

Here π_n *is the canonical projection* (see [10] p.57-61).

Proof. Let $x \in \mathcal{L}((A))$ with $\langle x, 1\rangle = 0$. Then there are $x_k \in \mathcal{L}(A)_k$, $k = 1, 2, \ldots$, such that $x = \sum_{k=0}^{\infty} x_k$. Then we see that

$$\exp(x) = 1 + \sum_{\ell=1}^{\infty} \frac{1}{(\ell!)^2} \sum_{k_1,\ldots k_\ell} \sum_{\sigma \in S_\ell} x_{k_{\sigma(1)}} \cdots x_{k_{\sigma(\ell)}}.$$

One can easily see that

$$j_m(\pi_n(\sum_{\sigma \in S_\ell} x_{k_{\sigma(1)}} \cdots x_{k_{\sigma(\ell)}})) = \pi_n(j_m(\sum_{\sigma \in S_\ell} x_{k_{\sigma(1)}} \cdots x_{k_{\sigma(\ell)}})).$$

So we have our assertion. □

Let $E_m = \mathcal{L}(A)_{\leq m} \cap (\sum_{u \in A^{**}} \mathbf{R}u)$, $m \geq 1$, and let $\Phi_m : E_m \to \mathbf{R}\langle A\rangle_{\leq m}$, $m \geq 2$, be an algebraic map given by

$$\Phi_m(x) = j_m(\exp(x + v_0)), \qquad x \in E_m.$$

Then by Proposition 7, we see that

$$\pi_1(\Phi_m(x)) = x + v_0, \qquad x \in E_m.$$

So we see that Φ_m is an immersion and $\Phi_m(E_m)$ is a closed manifold in $\mathbf{R}\langle A\rangle_{\leq m}$ of dimensions $dim\ E_m$.

Lemma 8. *The distribution of $j_m(\log X(1) - v_0)$ on E_m is absolutely continuous and its density is smooth for any $m \geq 2$.*

Proof. This lemma is somehow well-known in Malliavin calculus, so we give a sketch of a proof only. Let $Y = j_m(\log X(1) - v_0)$. Let H be the Cameron-Martin space of d-dimensional Wiener process, that is, H is the Hilbert space consisting of $h = (h^1, \dots, h^d) : [0, \infty) \to \mathbf{R}$ such that $h^i(t)$, $i = 1, \dots, d$, are absolutely continuous in t, and

$$\| h \|_H^2 = \sum_{i=1}^d \int_0^\infty |\frac{d}{dt} h^i(t)|^2 dt < \infty.$$

Then we see that for each $h \in H$

$$D(X(t))(h) = \sum_{i=0}^d \int_0^t D(X(s))(h) v_i \circ dB^i(s) + \sum_{i=1}^d \int_0^t X(s) v_i \frac{d}{ds} h^i(s) ds,$$

and so we have

$$D(X(t))(h) X(t)^{-1} = \sum_{i=1}^d \int_0^t X(s) v_i X(s)^{-1} \frac{d}{ds} h^i(s) ds, \qquad t \geq 0.$$

Note that for $w \in \mathbf{R}\langle A\rangle$

$$X(t) w X(t)^{-1} = w + \sum_{i=0}^d \int_0^t X(s)[v_i, w] X(s)^{-1} \circ dB^i(s), \qquad t \geq 0.$$

Then we have

$$j_m(D(X(T))(h) X(T)^{-1}) = \sum_{i=1}^d \int_0^T (\sum_{u \in \mathbf{R}\langle A\rangle_{\leq m-1}} B(t; u) r(uv_i)) \frac{d}{ds} h^i(t) dt,$$
$$T \geq 0.$$

Here r is an operator defined in [10] p.20 . Then by the usual argument (e.g. [4]), we see that

$$E[\inf\{\| \langle j_m(D(X(1))(\cdot) X(1)^{-1}), w\rangle \|_{H^*};\ w \in E_m, \langle w, w\rangle = 1\}^{-p}] < \infty,$$
$$p \in (1, \infty).$$

Note that $j_m(X(1)) = \Phi_m(Y)$. So we have our assertion from Taniguchi [11]. □

4. Proof of Theorem 3

For any vector field $V \in C_b^\infty(\mathbf{R}^N;\mathbf{R}^N)$ on $\mathbf{R}^N$, let us think of ODE given by

$$\frac{d}{dt}x(t,x) = V(x(t,x)), \qquad t > 0,$$
$$x(0,x) = x \in \mathbf{R}^N,$$

and let us define a diffeomorphism $\exp(V) : \mathbf{R}^N \to \mathbf{R}^N$ by $\exp(V)(x) = x(1,x)$. Then we have

$$\frac{d}{dt}f(\exp(tV)(x)) = (Vf)(\exp(tV)(x))$$

for any $f \in C^\infty(\mathbf{R}^N)$.

So we have the following.

Proposition 9. *For any vector field $V \in C_b^\infty(\mathbf{R}^N;\mathbf{R}^N)$,*

$$f(\exp(tV)(x)) = \sum_{k=0}^{n} \frac{t^k}{k!}(V^k f)(x) + \int_0^t \frac{(t-s)^n}{n!}(V^{n+1}f)(\exp(sV)(x))ds,$$

for any $n \geq 1, t > 0, x \in \mathbf{R}^N$ and $f \in C^\infty(\mathbf{R}^N)$. In particular,

$$|f(\exp(V)(x)) - \sum_{k=0}^{n} \frac{1}{k!}(V^k f)(x)| \leq \frac{1}{(n+1)!} \parallel V^n f \parallel_\infty,$$

for any $n \geq 1, x \in \mathbf{R}^N$ and $f \in C^\infty(\mathbf{R}^N)$.

Corollary 10. *Let $z \in \mathcal{L}((A))$ and $n, m \geq 1$. Then we have*

$$|f(\exp(\Phi(j_m z))(x)) - \sum_{k=0}^{n} \frac{1}{k!}(\Phi((j_m z)^k)f)(x)|$$
$$\leq \frac{1}{(n+1)!} \parallel \Phi((j_m z)^{n+1})f \parallel_\infty,$$

for any $x \in \mathbf{R}^N$ and $f \in C^\infty(\mathbf{R}^N)$.

Then we have the following.

Lemma 11. *Let $z_1, z_2 \in \mathcal{L}((A))$ and $m \geq 1$. Then we have*

$$|f(\exp(\Phi(j_m z_1))(\exp(\Phi(j_m z_2))(x)) - \sum_{k+\ell \leq m} \frac{1}{k!\ell!}(\Phi((j_m z_2)^k (j_m z_1)^\ell f)(x)|$$

$$\leq \sum_{\ell=0}^{m} \frac{1}{\ell!(m+1-\ell)!} \parallel \Phi((j_m z)^{m+1-\ell}(j_m z_1)^\ell)f \parallel_\infty,$$

for any $x \in \mathbf{R}^N$ and $f \in C^\infty(\mathbf{R}^N)$.

Proof. Note that

$$|f(\exp(\Phi(j_m z_1))(x)) - \sum_{\ell=0}^{m} \frac{1}{\ell!}(\Phi((j_m z_1)^\ell)f)(x)| \leq \frac{1}{(m+1)!} \parallel \Phi((j_m z_1)^{m+1})f \parallel_\infty,$$

and

$$|(\Phi((j_m z_1)^\ell)f)(\exp(\Phi j_m z_2)(x)) - \sum_{k=0}^{m-\ell} \frac{1}{k!}(\Phi((j_m z_2)^k (j_m z_1)^\ell)f)(x)|$$

$$\leq \frac{1}{(m+1-\ell)!} \parallel \Phi((j_m z_2)^{m+1-\ell}(j_m z_1)^\ell)f \parallel_\infty .$$

Thus we have our assertion. □

Corollary 12. *Let* $z_1, z_2 \in \mathcal{L}((A))$ *and* $m \geq 1$. *Then we have*

$$|f(\exp(\Phi(j_m z_1))(\exp(\Phi(j_m z_2))(x)) - (\Phi(j_m(\exp(j_m z_2)\exp(j_m z_1)))f)(x)|$$

$$\leq \sum_{2 \leq k+\ell \leq m+1} \frac{1}{\ell!k!} \parallel \Phi((j_{m^{m+1}} - j_m)((j_m z_2)^k (j_m z_1)^\ell))f \parallel_\infty,$$

for any $x \in \mathbf{R}^N$ *and* $f \in C^\infty(\mathbf{R}^N)$.

Proof. Note that

$$(j_m z_2)^k (j_m z_1)^\ell = j_{m^{m+1}}((j_m z_2)^k (j_m z_1)^\ell),$$

if $k + \ell \leq m + 1$. So we have

$$j_m(\exp(j_m z_2)\exp(j_m z_1))$$

$$= \sum_{k+\ell \leq m} \frac{1}{k!\ell!}(j_m z_2)^k (j_m z_1)^\ell - \sum_{2 \leq k+\ell \leq m} \frac{1}{\ell!k!}(j_{m^{m+1}} - j_m)((j_m z_2)^k (j_m z_1)^\ell),$$

and

$$\sum_{\ell=0}^{m} \frac{1}{\ell!(m+1-\ell)!}(j_m z)^{(m+1-\ell)}(j_m z_1)^\ell = \sum_{k+\ell=m+1} \frac{1}{\ell!k!}(j_{m^{m+1}} - j_m)((j_m z_2)^k (j_m z_1)^\ell).$$

So we have our assertion by Lemma 11. □

Lemma 13. *For any $n \geq 1$, there is a $C_n > 0$ such that*

$$\| \Phi(j_n z) f \|_\infty \leq C_n \| j_n z \|_2 \| \nabla f \|_{C^{n-1}}$$

for any $z \in \mathcal{L}((A))$ and $f \in C^\infty(\mathbf{R}^N)$. Here

$$\| f \|_{C^n} = \| f \|_\infty + \sum_{k=1}^{n} \sum_{\alpha_1,\ldots,\alpha_k=1}^{N} \| \frac{\partial^k}{\partial x^{\alpha_1} \cdots \partial x^{\alpha_k}} f \|_\infty, \quad n \geq 0.$$

Proof. For each $w \in A^* \setminus \{1\}$, there exists a $C_w > 0$ such that

$$\| \Phi(\iota(w)) f \|_\infty \leq C_w \| \nabla f \|_{C^{|w|-1}}$$

for any $f \in C^\infty(\mathbf{R}^N)$. Then we have

$$\| \Phi(j_n z) f \|_\infty \leq \sum_{w \in A, 1 \leq \|w\| \leq n} C_w |\langle z, \iota(w) \rangle| \| \nabla f \|_\infty .$$

This implies our assertion. □

Lemma 14. *For any $m \geq 1$, there is a $C_m > 0$ such that*

$$|f(\exp(\Phi(j_m \Psi_s z_1))(\exp(\Phi(j_m \Psi_s z_2))(x))$$
$$- f(\exp(\Phi(j_m(\log(\exp(j_m \Psi_s z_2)\exp(j_m \Psi_s z_1))))(x))|$$
$$\leq C_m s^{(m+1)/2}(1 + \| j_m z_1 \|_2 + \| j_m z_2 \|_2)^{m^2(m+1)} \| \nabla f \|_{C^m}$$

for any $s \in (0,1]$, $z_1, z_2 \in \mathcal{L}((A))$ and $f \in C^\infty(\mathbf{R}^N)$.

Proof. Let $w = \log(\exp(j_m z_2)\exp(j_m z_1))$. Then we have

$$\Psi_s w = \log(\exp(j_m \Psi_s z_2)\exp(j_m \Psi_s z_1))$$

and

$$j_m \exp(j_m \Psi_s w) = j_m(\exp(j_m \Psi_s z_2)\exp(j_m \Psi_s z_1)).$$

Then letting $z_1 = w$ and $z_2 = 0$ in Corollary 12, we have

$$|f(\exp(\Phi(j_m \Psi_s w))(x)) - (\Phi(j_m(\exp(j_m \Psi_s w)))f)(x)|$$

$$\leq \sum_{k=2}^{m+1} \frac{1}{k!} \| \Phi((j_{m^{m+1}} - j_m)((j_m w)^k)) f \|_\infty .$$

Therefore by Corollary 12, there is a $C > 0$

$$|f(\exp(\Phi(j_m \Psi_s z_1))(\exp(\Phi(j_m \Psi_s z_2))(x)) - f(\exp(\Phi(j_m \Psi_s w))(x))|$$

$$\le C\Big(\sum_{2\le k+\ell\le m+1} \| (j_{m^{m+1}} - j_m)((j_m\Psi_s z_2)^k (j_m\Psi_s z_1)^\ell) \|_2$$
$$+ \sum_{k=2}^{m+1} \| (j_{m^{m+1}} - j_m)((j_m w)^k)) \|_2\Big) \| \nabla f \|_{C^m}$$

for any $s \in (0,1]$ and $f \in C^\infty(\mathbf{R}^N)$. Note that

$$\| (j_{m^{m+1}} - j_m)((j_m\Psi_s z_2)^k (j_m\Psi_s z_1)^\ell) \|_2 \le s^{(m+1)/2} \| (j_m z_2)^k (j_m z_1)^\ell \|_2$$
$$\le s^{(m+1)/2} \| j_m z_2 \|_2^k \| j_m z_1 \|_2^\ell$$

and that

$$\| j_m w \|_2 = \| j_m\Big(\sum_{i=1}^{m} \frac{(-1)^{i-1}}{i}\Big(\sum_{1\le k+\ell\le m} \frac{1}{k!\ell!}(j_m z_2)^k (j_m z_1)^\ell)\Big)^i\Big) \|_2$$
$$\le \sum_{i=1}^{m}\Big(\sum_{1\le k+\ell\le m} \| j_m z_2 \|_2^k \| j_m z_1 \|_2^\ell\Big)^i.$$

These imply our assertion. □

Corollary 15. *Let ξ_1, ξ_2 be $\mathcal{L}((A))$-valued $L^{\infty-}$ random variable. Then for any $m \ge 1$ and $p \in [1,\infty)$, there is a $C > 0$ such that*

$$\| \exp(\Phi(j_m\Psi_s\xi_1))(\exp(\Phi(j_m\Psi_s\xi_2))(x))$$
$$- \exp(\Phi(j_m(\log(\exp(j_m\Psi_s\xi_2)\exp(j_m\Psi_s\xi_1))))(x) \|_{L^p} \le Cs^{(m+1)/2}$$

for any $s \in (0,1]$ and $x \in \mathbf{R}^N$.

Proof. Let $f(x) = x^i$, $x = (x^1, \ldots, x^N) \in \mathbf{R}^N$. Then we have $\| \nabla f \|_{C^n} = 1$. Applying Lemma 14, we have our assertion. □

Proposition 16. (1) *For any $m \ge 1$ and $f \in C^\infty(\mathbf{R}^N)$,*

$$f(X(t,x))$$
$$= (\Phi(j_m X(t))f)(x) + \sum{}' \int_0^t \circ dB^{i_1}(s_1) \int_0^{s_1} \circ dB^{i_2}(s_2) \cdots$$
$$\int_0^{s_{n-1}} \circ dB^{i_n}(s_n)(V_{i_n} \cdots V_{i_1} f)(X(s_n,x)).$$

Here $\sum'$ is the summation taken for $i_1, \ldots, i_n = 0, 1, \ldots, N$ such that $\| v_{i_{n-1}} v_{i_{n-2}} \cdots v_{i_1} \| \le m$ and $\| v_{i_n} v_{i_{n-1}} \cdots v_{i_1} \| \ge m+1$.
(2) *For any $m \ge 1$ and $p \in [1,\infty)$, there is a $C > 0$ such that*

$$\| f(X(t,x)) - (\Phi(j_m(X(t)))f)(x) \|_{L^p} \le Ct^{(m+1)/2} \| \nabla f \|_{C^{m+1}}$$

for any $t \in (0,1]$ and $f \in C^\infty(\mathbf{R}^N)$.

Proof. The assertion (1) is easy to prove by induction in m. The assertion (2) follows from the fact that

$$\int_0^{s_{n-1}} \circ dB^{i_n}(s_n)(V_{i_n}\dots V_{i_1}f)(X(s_n,x))$$
$$= \int_0^{s_{n-1}} (V_{i_n}\dots V_{i_1}f)(X(s_n,x))dB^{i_n}(s_n)$$
$$+\frac{1}{2}\sum_{j=1}^{N}\delta_{ji_n}\int_0^{s_{n-1}}(V_jV_{i_n}\dots V_{i_1}f)(X(s_n,x))ds_n.$$

This completes the proof. □

Corollary 17. *For any $m \geq 1$, there is a $C > 0$ such that*

$$|E[f(X(s,x))] - E[f(\exp(\Phi(j_m\Psi_s \log X(1))))(x)]| \leq Cs^{(m+1)/2} \parallel \nabla f \parallel_\infty$$

for any $x \in \mathbf{R}^N$, $s \in (0,1]$ and $f \in C_b^\infty(\mathbf{R}^N)$.

Proof. Let $H(x) = x$, $x \in \mathbf{R}^N$. Then by Proposition 16 (2), there is a $C_1 > 0$ such that

$$\parallel X(s,x) - (\Phi(j_m(X(s)))H)(x) \parallel_{L^1} \leq C_1 s^{(m+1)/2}, \qquad x \in \mathbf{R}^N,\ s \in (0,1].$$

So we see that

$$|E[f(X(s,x))] - E[f((\Phi(j_m(X(s)))H)(x))]| \leq C_1 s^{(m+1)/2} \parallel \nabla f \parallel_\infty,$$
$$x \in \mathbf{R}^N,\ s \in (0,1].$$

Also by Corollary 12 we have

$$|\exp(\Phi(j_m\Psi_s \log X(1)))(x) - (\Phi(j_m(\Psi_s X(1)))H)(x)|$$
$$\leq \sum_{k=2}^{m+1} \frac{1}{k!} s^{(m+1)/2} \parallel \Phi((j_{m^{m+1}} - j_m)((j_m \log X(1))^k))H \parallel_\infty,$$
$$x \in \mathbf{R}^N,\ s \in (0,1].$$

So we see that there is a $C_2 > 0$ such that

$$\parallel \exp(\Phi(j_m\Psi_s \log X(1)))(x) - (\Phi(j_m(\Psi_s X(1)))H)(x) \parallel_{L^1} \leq C_2 s^{(m+1)/2}$$

for any $x \in \mathbf{R}^N$, $s \in (0,1]$, which implies that

$$|E[f(\exp(\Phi(j_m\Psi_s \log X(1)))(x))] - E[f((\Phi(j_m(\Psi_s X(1)))H)(x))]|$$
$$\leq C_2 s^{(m+1)/2} \parallel \nabla f \parallel_\infty$$

for any $x \in \mathbf{R}^N$, $s \in (0,1]$. Since $j_m(X(s))$ and $j_m\Psi_s X(1)$ has the same law, we have our assertion. □

Lemma 18. *Let $m \geq 2$ and ξ is a m-$\mathcal{L}$-similar $\mathcal{L}((A))$-valued random variable. Then there is a constant $C > 0$ such that*

$$|E[f(X(s,x))] - E[f(\exp(\Phi(j_m \Psi_s \xi))(x))]|$$

$$\leq C\left(\sum_{k=m+1}^{m^{m+1}} s^{k/2} \| f \|_{V,k} + s^{(m+1)/2} \| \nabla f \|_\infty\right)$$

for any $s \in (0,1]$ and $f \in C_b^\infty(\mathbf{R}^N)$.

Proof. Let $\eta_0 = \log(\exp(-v_0)X(1))$ and $\eta_1 = \log(\exp(-v_0)\exp(\xi))$. Then η_0 and η_1 are $\mathcal{L}((A))$-valued $L^{\infty-}$ random variable and we see that

$$E[j_m(\exp(\eta_0))] = E[j_m(\exp(-v_0)j_m(X(1)))]$$

$$= E[j_m(\exp(-v_0)j_m(\exp(\xi))] = E[j_m(\exp(\eta_1))].$$

Note that $j_m(\eta_i) \in \mathcal{L}(A) \cap (\sum_{w \in A^{**}} \mathbf{R}w)$, $i = 0,1$. So there is a $C_1 > 0$ such that

$$\| \Phi((j_{m^{m+1}} - j_m)(j_m \Psi_s \eta_i)^\ell) f \|_\infty \leq C_1 \sum_{k=m+1}^{m^{m+1}} s^{k/2} \| f \|_{V,k}$$

for any $i = 0,1$, $s \in (0,1]$ and $f \in C_b^\infty(\mathbf{R}^N)$. So we see that there is a $C_2 > 0$ such that

$$\| f(\exp(\Phi(j_m \Psi_s \eta_i))(y)) - (\Phi(j_m(\exp(j_m \Psi_s \eta_i))f)(y) \|_{L^1}$$

$$\leq C_2 \sum_{k=m+1}^{m^{m+1}} s^{k/2} \| f \|_{V,k}$$

for any $i = 0,1$, $s \in (0,1]$, $y \in \mathbf{R}^N$, and $f \in C_b^\infty(\mathbf{R}^N)$. However, $E[\Phi(j_m(\exp(j_m \Psi_s \eta_i))f)(y)]$, $i = 0,1$ are coincident. So letting $y = \exp(\Phi(j_m \Psi_s v_0))(x)$, we have

$$|E[f(\exp(\Phi(j_m \Psi_s \eta_0))(\exp(\Phi(j_m \Psi_s v_0))(x)))]$$
$$- E[f(\exp(\Phi(j_m \Psi_s \eta_1))(\exp(\Phi(j_m \Psi_s v_0))(x)))]|$$

$$\leq 2C_2 \sum_{k=m+1}^{m^{m+1}} s^{k/2} \| f \|_{V,k}$$

for any $x \in \mathbf{R}^N$, and $f \in C_b^\infty(\mathbf{R}^N)$. Note that

$$j_m \log(\exp((j_m v_0))(\exp(j_m \eta_i))) = j_m \log(\exp(v_0)(\exp(\eta_i))), \qquad i = 0,1.$$

Then by Corollaries 15 and 17, we have our assertion.

This completes the proof. □

Now Theorem 3 follows from Lemma 18, since

$$|E[f(\exp(\Phi(j_m\Psi_s\xi))(x))]| - E[f(\exp(\Phi(j_m\Psi_s\xi))(x) + Y(s,x))]| \\ \leq E[|Y(s,x)|] \parallel \nabla f \parallel_\infty .$$

This completes the proof of Theorem 3.

References

1. Bally, D., Talay, D.: The law of the Euler scheme for stochastic differential equations I. Convergence rate of the distribution function, Probab. Theory Relat. Fields **104**, 43-60 (1996)
2. Kloeden, P.E., Platen, E.: Numerical Solution of Stochastic Differential Equations. Applications of Mathematics vol.**23**, Springer, Berlin 1994
3. Kusuoka, S.: Approximation of expectation of diffusion processes and mathematical finance. In: Advanced Studies in Pure Mathematics **31**, Proceedings of Final Taniguchi Symposium, Nara 1998 (edited by Sunada, T.), Mathematical Society of Japan, pp.147-165, 2001
4. Kusuoka, S.: Malliavin calculus revisited. J. Math. Sci. Univ. Tokyo **10**, 261-277 (2003)
5. Kusuoka, S., Stroock, D.W.: Applications of Malliavin calculus II, J. Fac. Sci. Univ. Tokyo Sect. IA Math. **32**, 1-76 (1985)
6. Kusuoka, S., Stroock, D.W.: Applications of Malliavin calculus III, J. Fac. Sci. Univ. Tokyo Sect. IA Math. **34**, 391-442 (1987)
7. Lyons, T., Victoir, N.: Cubature on Wiener Space (Preprint)
8. Ninomiya, S.: A new simulation scheme of diffusion processes: application of the Kusuoka approximation to finance problems. Mathematics and Computer in Simulation **62**, 479-486 (2003)
9. Ninomiya, S.: A partial sampling method applied to the Kusuoka approximation', Monte Carlo Methods and Applications **9**, 27-38 (2003)
10. Reutenauer, C.: Free Lie Algebras. Clarendon Press, Oxford 1993
11. Taniguchi, S.: Malliavin's stochastic calculus of variations for manifold-valued Wiener functionnals and its applications. Z.Wahrsch. verw. Gebiete **65**, 269-290 (1983)

Adv. Math. Econ. 6, 85–122 (2004)

Advances in
MATHEMATICAL
ECONOMICS

Optimal solutions of the Monge problem

Vladimir L. Levin*

Central Economics and Mathematics Institute of Russian Academy of Sciences, 47 Nakhimovskii Prospect, 117418 Moscow, Russia
(e-mail: vllevin@mail.ru)

Received: April 4, 2003
Revised: May 21, 2003

JEL classification: C65

Mathematics Subject Classification (2000): 49Q20, 28A35

Summary. We obtain optimality conditions for Monge solutions of the Monge—Kantorovich problem with a smooth cost function. Also we give explicit solutions to Monge problems and to Monge—Kantorovich problems for several natural classes of cost functions.

Key words: Borel measure, marginal measure, measure preserving map, Monge problem, Monge–Kantorovich problem.

1. Introduction

In this article, X and Y are closed domains in spaces $\mathbb{R}^n$ and $\mathbb{R}^m$, σ_1 and σ_2 are positive Borel measures on them, $\sigma_1 X = \sigma_2 Y$, and $c : X \times Y \to \mathbb{R}$ is a bounded continuous cost function. The *Monge problem* $MP(c; \sigma_1, \sigma_2)$ is to minimize the functional

$$\mathcal{F}(f) := \int_X c(x, f(x))\, \sigma_1(dx), \tag{1.1}$$

over the set $\Phi(\sigma_1, \sigma_2)$ of measure-preserving Borel maps $f : X \to Y$. A map f is called *measure-preserving* if $f(\sigma_1) = \sigma_2$, that is $\sigma_2 B_Y = \sigma_1 f^{-1}(B_Y)$ for every Borel set $B_Y \subset Y$. The optimal value of the Monge problem is thus

$$\mathcal{V}(c; \sigma_1, \sigma_2) := \inf\{\mathcal{F}(f) : f \in \Phi(\sigma_1, \sigma_2)\}. \tag{1.2}$$

* Supported in part by Russian Foundation for Humanitarian Sciences (projects 01-02-00481, 03-02-00027).

The *Monge–Kantorovich problem* $MKP(c;\sigma_1,\sigma_2)$ with fixed marginal measures σ_1,σ_2 and a cost function c is a relaxation of the above Monge problem. It is an infinite linear program that has the form of minimizing the functional

$$\langle c,\mu\rangle := \int_{X\times Y} c(x,y)\,\mu(d(x,y)) \tag{1.3}$$

over the set $\Gamma(\sigma_1,\sigma_2)$ of positive Borel measures μ on $X\times Y$ satisfying $\pi_1\mu = \sigma_1$, $\pi_2\mu = \sigma_2$. Here, π_1,π_2 are the natural projecting maps of $X\times Y$ onto X, Y, while $\pi_1\mu$ and $\pi_2\mu$ are the corresponding marginal measures: for any Borel sets $B_X\subset X$ and $B_Y\subset Y$,

$$(\pi_1\mu)B_X : = \mu\pi_1^{-1}(B_X) = \mu(B_X\times Y),$$
$$(\pi_2\mu)B_Y : = \mu\pi_2^{-1}(B_Y) = \mu(X\times B_Y).$$

The optimal value of $MKP(c;\sigma_1,\sigma_2)$ is denoted as $\mathcal{C}(c;\sigma_1,\sigma_2)$ so that

$$\mathcal{C}(c;\sigma_1,\sigma_2) := \inf\{\langle c,\mu\rangle : \mu\in\Gamma(\sigma_1,\sigma_2)\}. \tag{1.4}$$

Each measure-preserving map $f\in\Phi(\sigma_1,\sigma_2)$ is associated with a measure $\mu_f = \mu_f(\sigma_1)\in\Gamma(\sigma_1,\sigma_2)$, where $\mu_f = (\mathrm{id}_X\times f)(\sigma_1)$. That is, for every Borel set $B\subset X\times Y$,

$$\mu_f B = \sigma_1(\mathrm{id}_X\times f)^{-1}(B) = \sigma_1\{x : (x,f(x))\in B\}. \tag{1.5}$$

Therefore, for every Borel set $B_X\subset X$, one has

$$\mu_f(B_X\times Y) = \sigma_1 B_X,$$

and for every Borel set $B_Y\subset Y$, one has

$$\mu_f(X\times B_Y) = \sigma_1\{x\in X : f(x)\in B_Y\} = \sigma_1 f^{-1}(B_Y) = \sigma_2 B_Y.$$

We get $\pi_1\mu_f = \sigma_1$ and $\pi_2\mu_f = \sigma_2$, so that μ_f actually belongs to $\Gamma(\sigma_1,\sigma_2)$. It is clear that $\langle c,\mu_f\rangle = \mathcal{F}(f)$, which implies

$$\mathcal{C}(c;\sigma_1,\sigma_2) \le \mathcal{V}(c;\sigma_1,\sigma_2). \tag{1.6}$$

Measures $\mu\in\Gamma(\sigma_1,\sigma_2)$ are called (feasible) *solutions* and measures of the form μ_f with $f\in\Phi(\sigma_1,\sigma_2)$ are called *Monge solutions* to $MKP(c;\sigma_1,\sigma_2)$. If there exists an optimal solution to $MKP(c;\sigma_1,\sigma_2)$, which is a Monge solution μ_f, then f is an optimal solution to $MP(c;\sigma_1,\sigma_2)$ and $\mathcal{C}(c;\sigma_1,\sigma_2) = \mathcal{V}(c;\sigma_1,\sigma_2)$. This is an immediate consequence of the identity $\langle c,\mu_f\rangle = \mathcal{F}(f)$.

In general case, inequality (1.6) is strict. We give two examples. In the first example, $\Phi(\sigma_1,\sigma_2)$ is empty at all, while in the second there exists an optimal solution f to $MP(c;\sigma_1,\sigma_2)$ but μ_f is not an optimal solution to $MKP(c;\sigma_1,\sigma_2)$.

Example 1.1. Take $\sigma_1 = \delta_{x^1}, \sigma_2 = \frac{1}{2}(\delta_{y^1} + \delta_{y^2}), y^1 \neq y^2$, where δ_z is the unit mass at the point z. Clearly, $\Phi(\sigma_1, \sigma_2)$ is empty, therefore $\mathcal{V}(c; \sigma_1, \sigma_2) = +\infty$. At the same time, $\mathcal{C}(c; \sigma_1, \sigma_2) = \frac{1}{2}(c(x^1, y^1) + c(x^1, y^2)) < +\infty$.

Example 1.2. $X = Y = [0, 1]$, $c(x, y) = \operatorname{dist}((x, y), M)$, $\sigma_1 = \frac{1}{2}\delta_0 + \lambda$, $\sigma_2 = \frac{1}{2}\delta_1 + \lambda$, where λ is the Lebesgue measure on $[1/4, 3/4]$,

$$M = (\{0\} \times [1/4, 3/4]) \cup ([1/4, 3/4] \times \{1\}) \cup \{(x, x) : 1/4 \leq x \leq 3/4\}.$$

We have $f(0) = 1 \quad \forall f \in \Phi(\sigma_1, \sigma_2)$; consequently, $\mathcal{V}(c; \sigma_1, \sigma_2) \geq c(0, 1)/2 = 1/8$. Since c is non-negative, the map f, $f(0) = 1$, $f(x) = x \quad \forall x, x \neq 0$, is an optimal solution to $MP(c; \sigma_1, \sigma_2)$ and $\mathcal{V}(c; \sigma_1, \sigma_2) = 1/8$. At the same time, $\langle c, \mu \rangle = 0$ for $\mu = \frac{1}{2}(\delta_0 \otimes \lambda + \lambda \otimes \delta_1)$, therefore μ is an optimal solution to $MKP(c; \sigma_1, \sigma_2)$ and $\mathcal{C}(c; \sigma_1, \sigma_2) = 0$.

Of a particular interest for us is the *classic* Monge problem, where $m = n$, X and Y have the same volume, and σ_1 and σ_2 are the n-dimensional Lebesgue measures on X and Y. The classic Monge problem with the Euclidean distance as the cost function was posed by Monge (1781); the statement of the Monge–Kantorovich problem is due to Kantorovich, who examined the case where $X = Y$ is an arbitrary metric compact space and c is the corresponding distance (see Kantorovich (1942), Kantorovich and Rubinshtein (1957, 1958), Kantorovich and Akilov (1984)). In such a case, the Monge–Kantorovich problem (MKP) with given marginals is equivalent to another variant of the Monge–Kantorovich problem, the MKP with a given marginal difference; see Kantorovich and Rubinshtein (1958), Kantorovich and Akilov (1984). Given a (signed) Borel measure $\rho (= \sigma_1 - \sigma_2)$ on X such that $\rho X = 0$, the MKP with a given marginal difference is to find the optimal value

$$\mathcal{A}(c; \rho) := \inf\{\langle c, \mu \rangle : \mu \geq 0, (\pi_1 - \pi_2)\mu = \rho\}, \tag{1.7}$$

where

$$\langle c, \mu \rangle = \int_{X \times X} c(x, y) \mu(d(x, y)).$$

The equivalence of two problems means that the equality $\mathcal{C}(c; \sigma_1, \sigma_2) = \mathcal{A}(c; \rho)$ is true, hence there exists an optimal solution μ to (1.7) such that $\pi_1 \mu = \sigma_1$, $\pi_2 \mu = \sigma_2$. The equality $\mathcal{C}(c; \sigma_1, \sigma_2) = \mathcal{A}(c; \rho)$ is extended to the case where c is a continuous (or merely a lower semi-continuous) function on $X \times X$ that vanishes on the diagonal ($c(x, x) = 0 \quad \forall x \in X$) and satisfies the triangle inequality ($c(x, y) \leq c(x, z) + c(z, y)$ whenever $x, y, z \in X$); see Levin and Milyutin (1979). (This equivalence fails to be true when the cost function does not satisfy the triangle inequality; see Levin (1990a).) The next result is a direct consequence of that equivalence.

Suppose that $X = Y \subset \mathbb{R}^n$ is compact, $c : X \times X \to \mathbb{R}$ is a continuous function vanishing on the diagonal and satisfying the triangle inequality, σ_1 and σ_2 are positive Borel measures on X, and $\sigma_1 X = \sigma_2 X$. We take the (signed) measure $\rho = \sigma_1 - \sigma_2$ on X and consider its Jordan decomposition $\rho = \rho_+ - \rho_-$, $\rho_+ := \rho \vee 0$, $\rho_- := (-\rho) \vee 0$. Here and below, $\vee$ and $\wedge$ stand for supremum and infimum in the Banach lattice $C(X)^*$ of finite Borel measures on X. A positive measure $\sigma_0 = \sigma_1 \wedge \sigma_2$ is defined then on X, and $\sigma_0 = \sigma_1 - \rho_+ = \sigma_2 - \rho_-$. We associate σ_0 with the measure $\mu_{\mathrm{id}_X}(\sigma_0)$ on $X \times X$ (cf. (1.5)), where id_X is the identical map on X, so that for every Borel set $B \subset X \times X$,

$$\mu_{\mathrm{id}_X}(\sigma_0)B = \sigma_0\{x : (x, x) \in B\}.$$

Theorem 1.1. *If μ' is an optimal solution to $MKP(c; \rho_+, \rho_-)$, then $\mu = \mu' + \mu_{\mathrm{id}_X}(\sigma_0)$ is an optimal solution to $MKP(c; \sigma_1, \sigma_2)$.*

Monge–Kantorovich problems with more general cost functions on (not necessarily metrizable) compact spaces are studied since 1974 (Levin (1974, 1975, 1977), Levin and Milyutin (1979)); for the cases of non-compact and non-topological spaces, see Levin (1990a, 1996, 1997a, 1997b, 1997c, 1999, 2001b). In these papers, duality theory is developed for both variants of the Monge–Kantorovich problem.

An important role in study of MKPs with a given marginal difference is played by the set

$$Q_0(c) := \{u \in \mathbb{R}^X : u(x^1) - u(x^2) \le c(x^1, x^2),\ x^1, x^2 \in X\}. \tag{1.8}$$

Notice that if c is continuous and vanishes on the diagonal, then $Q_0(c)$ consists of continuous functions, and if, in addition, c is a bounded function, then every $u \in Q_0(c)$ is bounded as well. In such a case, $Q_0(c)$ is precisely the constraint set of the infinite linear program dual to (1.7). The multifunction Q_0 arises in a natural way not only in mass transportation problems but also in such topics as cyclically monotone operators, dynamic optimization, approximation theory, and various parts of mathematical economics; see Levin (1990a, 1990b, 1991, 1997a, 1997c, 2001b), Carlier, Ekeland, Levin, and Shananin (2002). Many problems in those fields may be reduced to the single question of whether or not the set $Q_0(\zeta)$ is nonempty for some specific cost function ζ. If ζ is nonnegative, then $Q_0(\zeta)$ is nonempty since it contains constant functions. But if ζ is not assumed nonnegative, then the above question becomes nontrivial. The answer is given in terms of abstract cyclic monotonicity; see Levin (1996, 1999). The same question for some specific ζ determined by c and optimal $\mu \in \Gamma(\sigma_1, \sigma_2)$ proves to be crucial in study of $MKP(c; \sigma_1, \sigma_2)$ and of the corresponding Monge problem $MP(c; \sigma_1, \sigma_2)$; see Levin (2002, 2003) and Theorem 2.1 below. A different approach to study Monge–Kantorovich problems with given marginals and the corresponding Monge problems with cost functions $c(x, y) = h(x - y)$, where h is a smooth convex function, was proposed by Brenier (1987, 1991) and developed next by many authors; see Evans (1997) for references.

Sudakov (see Sudakov (1976)) was first who proved the existence of an optimal Monge solution to the Monge–Kantorovich problem; he considered the case where $c(x, y) = \|x - y\|$ (the norm is not necessarily Euclidean), σ_1 and σ_2 have compact supports and are absolutely continuous with respect to the n-dimensional Lebesgue measure. For such a cost function, the optimal solution is not unique. For $X \subset \mathbb{R}^n$ and a measure σ_1, which is absolutely continuous with respect to the n-dimensional Lebesgue measure, several cases are known today when a unique optimal solution to Monge–Kantorovich problem exists, which is the Monge solution; see Brenier (1987, 1991), Rüschendorf and Rachev (1990), Gangbo and McCann (1995, 1996), Levin (1998, 1999) (see also the book by Rachev and Rüschendorf (1998), where further references may be found). In most of publications, cost functions of the form $c(x, y) = h(x - y)$ were considered, cost functions of general form were studied in Levin (1998, 1999). The following existence and uniqueness results are particular cases of Theorems 6.1 and 6.2 in Levin (1999). Keeping in mind subsequent applications to classic Monge problems with convex X and Y (see Sections 5 and 6), we admit for the sake of simplicity that closed domains X and Y are bounded hence compact.

Recall that the *support* of a positive Borel measure σ on X is the minimum closed set of full measure σ; the support of σ is denoted as spt(σ).

We shall need two assumptions on σ_1 and c as follows:

(A_1) The measure σ_1 is absolutely continuous with respect to the n-dimensional Lebesgue measure.

(A_c) If, for $y^1, y^2 \in \text{spt}(\sigma_2)$, the functions $c(\cdot, y^1)$ and $c(\cdot, y^2)$ are differentiable at some point $x \in \text{int}X$ and their gradients at x coincide then $y^1 = y^2$.

The assumption (A_c) is satisfied, in particular, when $Y = X$ and $c(x, y) = x \cdot y = \sum_1^n x_i y_i$ or $c(x, y) = h(x - y)$ where h is strictly convex or strictly concave.

Theorem 1.2. *Let X be convex and $c : X \times Y \to \mathbb{R}$ continuous, and suppose (A_1), (A_c). Also suppose that all the functions $c(\cdot, y)$ $(y \in \text{spt}(\sigma_2))$ are convex. Then there exists a unique optimal solution to $MKP(c; \sigma_1, \sigma_2)$ and this optimal solution is the Monge solution $\mu_f = \mu_f(\sigma_1)$. Therefore, f is a unique, up to values on σ_1-negligible set, optimal solution to the Monge problem $MP(c; \sigma_1, \sigma_2)$.*

Theorem 1.3. *Let X be convex and $c : X \times Y \to \mathbb{R}$ continuous, and suppose (A_1), (A_c). Also suppose that all the functions $c(\cdot, y)$ $(y \in \text{spt}(\sigma_2))$ are concave. Then there exists a unique optimal solution to $MKP(c; \sigma_1, \sigma_2)$ and this optimal solution is the Monge solution $\mu_f = \mu_f(\sigma_1)$. Therefore, f is a* *unique, up to values on σ_1-negligible set, optimal solution to the Monge problem $MP(c; \sigma_1, \sigma_2)$.*

Theorem 1.4. *Let c be continuous, and suppose (A_1), (A_c). Also suppose that:*

(a) the set $\operatorname{int}\operatorname{spt}(\sigma_1)$ *is nonempty and its complement in* $\operatorname{spt}(\sigma_1)$ *is Lebesgue negligible;*

(b) the functions $c(\cdot, y)$ $(y \in \operatorname{spt}(\sigma_2))$ are differentiable on $\operatorname{int}\operatorname{spt}(\sigma_1)$ *and locally Lipschitz on* $\operatorname{spt}(\sigma_1)$ *uniformly with respect to y (the latter means that, for every $x \in \operatorname{spt}(\sigma_1)$, there exist a neighborhood V of x and a positive number $C = C(x)$ such that*

$$|c(x^1, y) - c(x^2, y)| \leq C\|x^1 - x^2\|$$

whenever $x^1, x^2 \in V$, $y \in \operatorname{spt}(\sigma_2)$). Then there exists a unique optimal solution to $MKP(c; \sigma_1, \sigma_2)$ and this optimal solution is the Monge solution $\mu_f = \mu_f(\sigma_1)$. Therefore, f is a unique, up to values on σ_1-negligible set, optimal solution to the Monge problem $MP(c; \sigma_1, \sigma_2)$.

Remark 1.1. In Theorems 1.2–1.4 (and in more general Theorems 6.1 and 6.2 in Levin (1999)), the uniqueness of optimal solutions is established for specific σ_1 and c. Unlike those individual uniqueness results, in Levin (2001a) we proved several generic uniqueness theorems, in which σ_1 and σ_2 with $\sigma_1 X = \sigma_2 Y$ are arbitrary. The simplest of them is as follows. Suppose that X and Y are bounded closed domains in Euclidean spaces, then the cost functions c, for which the optimal solution to $MKP(c; \sigma_1, \sigma_2)$ is unique, make up a massive (dense G_δ) subset in $C(X \times Y)$. However in this theorem, unique optimal solutions to the corresponding $MKP(c; \sigma_1, \sigma_2)$ need not be the Monge solutions.

The structure of the paper is as follows. In Section 2 (also see Levin (2002, 2003)), we prove that optimality of a measure $\mu \in \Gamma(\sigma_1, \sigma_2)$ is equivalent to nonemptiness of $Q_0(\varphi_{F_\mu})$ where

$$\varphi_{F_\mu}(x, z) = \inf_{y:(x,y) \in \operatorname{spt}(\mu)} (c(z, y) - c(x, y)),\ x, z \in Z := \pi_1 \operatorname{spt}(\mu).$$

As consequences of that equivalence, we obtain a similar equivalence result for a Monge solution μ_f to $MKP(c; \sigma_1, \sigma_2)$ and the function $\varphi^f(x, z) = c(z, f(x)) - c(x, f(x))$, where $f \in \Phi(\sigma_1, \sigma_2)$ is continuous, and some results on optimality for various measures μ and μ_f. In Section 3, conditions for $Q_0(\zeta)$ to be nonempty are given for a bounded smooth function that vanishes on the diagonal. Similar conditions were obtained earlier in Levin (1990b) for X being an *open* domain in $\mathbb{R}^n$ (see also Levin (1996) where those results were extended to X being an open domain in a locally convex space), and now we extend them to the case where X is a *closed* domain. In Section 4 (also see Levin (2002)), basing on results of Section 3 we characterize optimal Monge solutions to MKPs with smooth cost functions. In Sections 5 and 6, explicit solutions to Monge problems are studied. Section 5 is devoted to theory (most results of Section 5 are announced in Levin (2003)[1]), and in Section 6 we give some concrete examples.

2. Optimality conditions for solutions to MKPs in terms of $Q_0(\zeta)$

Given a Monge–Kantorovich problem $MKP(c; \sigma_1, \sigma_2)$, we consider the set of real-valued functions on X, $L := \{-c(\cdot, y) : y \in \text{spt}(\sigma_2)\}$, where spt stands for the support of the corresponding measure. Following Levin (1999), we associate every $\mu \in \Gamma(\sigma_1, \sigma_2)$ with the multifunction $F_\mu : X \to L$,

$$F_\mu(x) := \{-c(\cdot, y) : (x, y) \in \text{spt}(\mu)\}. \tag{2.1}$$

(F_μ is well–defined because the projection of spt(μ) onto Y is contained in spt(σ_2).) Let us denote $Z = \text{dom}F_\mu := \{x \in X : F_\mu(x) \neq \phi\}$ and define on $Z \times Z$ the function

$$\varphi_{F_\mu}(x, z) := \inf\{l(x) - l(z) : l \in F_\mu(x)\} = \inf_{y:(x,y)\in\text{spt}(\mu)} (c(z, y) - c(x, y)). \tag{2.2}$$

Theorem 2.1. *A measure $\mu \in \Gamma(\sigma_1, \sigma_2)$ is an optimal solution to $MKP(c; \sigma_1, \sigma_2)$ if and only if the set $Q_0(\varphi_{F_\mu})$ is nonempty.*

Proof. It follows from the equivalence (a)⇔(b) of Theorem 5.1 in Levin (1999) that μ is optimal if and only if the multifunction F_μ is L-cyclical monotone that is, for every positive integer p and for every cycle $x^1, \ldots, x^p, x^{p+1} = x^1$ in domF_μ, the inequality

$$\sum_{k=1}^{p} (l_k(x^k) - l_k(x^{k+1})) \geq 0$$

holds whenever $l_k = -c(\cdot, y^k) \in F_\mu(x^k)$, $k = 1, \ldots, p$. On the other hand, as follows from Theorem 2.1 in Levin (1999), F_μ is L-cyclical monotone if and only if $Q_0(\varphi_{F_\mu})$ is nonempty. □

[1] There is a disappointing defect in formulation of Theorem 4 in Levin (2003); a correct formulation is given in Theorem 5.2 below.

Corollary 2.1. *Suppose that μ is an optimal solution to $MKP(c;\sigma_1,\sigma_2)$, μ' is a positive Borel measure on $X \times Y$, and σ_1', σ_2' are its projections on X, Y. Then:*

(a) if $\mathrm{spt}(\mu') \subset \mathrm{spt}(\mu)$, *then μ' is an optimal solution to $MKP(c;\sigma_1',\sigma_2')$;*

(b) if μ is the unique optimal solution to $MKP(c;\sigma_1,\sigma_2)$ and $0 \le \mu' \le \mu$, then μ' is the unique optimal solution to $MKP(c;\sigma_1',\sigma_2')$.

Proof. (a) According to Theorem 2.1, there exists a function $u \in Q_0(\varphi_{F_\mu})$. Since $\mathrm{spt}(\mu') \subset \mathrm{spt}(\mu)$, one has $\mathrm{dom}F_{\mu'} \subseteq \mathrm{dom}F_\mu$ and $F_{\mu'}(x) \subseteq F_\mu(x)$ for every $x \in \mathrm{dom}F_{\mu'}$; therefore $\varphi_{F_{\mu'}}(x,z) \ge \varphi_{F_\mu}(x,z)$ whenever $x, z \in \mathrm{dom}F_{\mu'}$. It follows that the restriction of u to $Z' := \mathrm{dom}F_{\mu'}$ belongs to the set $Q_0(\varphi_{F_{\mu'}})$ hence this set is nonempty. Now applying Theorem 2.1 shows that μ' is an optimal solution to $MKP(c;\sigma_1',\sigma_2')$.

(b) Since $\mu-\mu' \ge 0$, one has $\sigma_1-\sigma_1' = \pi_1(\mu-\mu') \ge 0, \sigma_2-\sigma_2' = \pi_2(\mu-\mu') \ge 0$. It follows from (a) that μ' is an optimal solution to $MKP(c;\sigma_1',\sigma_2')$ and $\mu-\mu'$ is an optimal solution to $MKP(c;\sigma_1-\sigma_1',\sigma_2-\sigma_2')$. Suppose now that μ'' is an arbitrary optimal solution to $MKP(c;\sigma_1',\sigma_2')$. Then one has

$$\langle c, \mu''\rangle = \mathcal{C}(c;\sigma_1',\sigma_2') = \langle c, \mu'\rangle.$$

We obtain

$$\mathcal{C}(c;\sigma_1,\sigma_2) = \langle c,\mu\rangle = \langle c,\mu\rangle - \langle c,\mu'\rangle + \langle c,\mu''\rangle = \langle c, \mu-\mu'+\mu''\rangle,$$

and as $\mu-\mu'+\mu'' \in \Gamma(\sigma_1,\sigma_2)$, it follows that $\mu-\mu'+\mu''$ is an optimal solution to $MKP(c;\sigma_1,\sigma_2)$. Since μ was supposed to be the unique optimal solution to $MKP(c;\sigma_1,\sigma_2)$, we get $\mu''=\mu'$, which completes the proof. □

Notice now that for a continuous map $f \in \Phi(\sigma_1,\sigma_2)$, one has

$$\mathrm{spt}(\mu_f(\sigma_1)) = (\mathrm{spt}(\sigma_1)\times Y) \cap \mathrm{gr}(f),$$

where $\mathrm{gr}(f)$ denotes the graph of f. In such a case, $F := F_{\mu_f(\sigma_1)}$ is single–valued on $Z = \mathrm{dom}F = \mathrm{spt}(\sigma_1)$, $F(x) = -c(\cdot, f(x))$ for every $x \in Z$, and φ_F becomes the restriction to $Z \times Z$ of the function φ on $X \times X$,

$$\varphi(x,z) = \varphi^f(x,z) := c(z,f(x)) - c(x,f(x)). \tag{2.3}$$

Corollary 2.2. *Suppose that $f \in \Phi(\sigma_1,\sigma_2)$ is continuous. Then $\mu_f(\sigma_1)$ is an optimal solution to $MKP(c;\sigma_1,\sigma_2)$ if and only if the set $Q_0(\varphi|_{Z\times Z})$ is nonempty.*

Proof. This is an immediate consequence of Theorem 2.1. □

Corollary 2.3. *If $f \in \Phi(\sigma_1,\sigma_2)$ is continuous and $\mu_f(\sigma_1)$ is an optimal solution to $MKP(c;\sigma_1,\sigma_2)$, then, for any positive measure σ on X with* $\mathrm{spt}(\sigma) \subset \mathrm{spt}(\sigma_1)$, *the measure $\mu_f(\sigma)$ is an optimal solution to $MKP(c;\sigma,f(\sigma))$.*

Proof. By Theorem 2.1, there exists a function $u \in Q_0(\varphi|_{Z\times Z})$ where φ is given by (2.3) and $Z = \text{spt}(\sigma_1)$. Then the restriction of u to $Z_\sigma := \text{spt}(\sigma)$ belongs to $Q_0(\varphi|_{Z_\sigma\times Z_\sigma})$, and the result follows from Theorem 2.1. □

Given a continuous map $f : X \to X$, we consider its iterations $f^1 = f$, $f^k = f \circ f^{k-1}$, $k = 2, 3, \ldots$.

Corollary 2.4. *Suppose that $X = Y$ and that the cost function c vanishes on the diagonal ($c(x, x) = 0 \quad \forall x \in X$) and satisfies the triangle inequality:*

$$c(x, y) \le c(x, z) + c(z, y) \quad \forall x, y, z \in X.$$

If $\text{spt}(\sigma) = X$ *and $\mu_f(\sigma)$ is an optimal solution to $MKP(c; \sigma, f(\sigma))$, then $\mu_g(\sigma)$ is an optimal solution to $MKP(c; \sigma, g(\sigma))$ whenever $g = f^k$, $k = 2, 3, \ldots$.*

Proof. By Corollary 2.2, there is a function $u \in Q_0(\varphi)$ so that

$$u(x) - u(y) \le c(y, f(x)) - c(x, f(x)) \text{ for all } x, y \in X. \tag{2.4}$$

Let us apply (2.4) to pairs $(x, f(x)), (f(x), f^2(x)), \ldots, (f^{k-1}(x), y)$ and take into account that c vanishes on the diagonal. We have

$$\begin{aligned} u(x) - u(f(x)) &\le -c(x, f(x)), \\ u(f(x)) - u(f^2(x)) &\le -c(f(x), f^2(x)), \\ &\ldots \\ u(f^{k-1}(x)) - u(y) &\le c(y, f^k(x)) - c(f^{k-1}(x), f^k(x)). \end{aligned}$$

Now summing up these inequalities yields

$$\begin{aligned} u(x) - u(y) \le -c(x, f(x)) - c(f(x), f^2(x)) - \\ \cdots - c(f^{k-1}(x), f^k(x)) + c(y, f^k(x)), \end{aligned}$$

and as c satisfies the triangle inequality, we get $u(x) - u(y) \le c(y, f^k(x)) - c(x, f^k(x))$. That is $u \in Q_0(\varphi^g)$, where $g = f^k$, and the proof is completed by applying Corollary 2.2. □

3. Conditions for $Q_0(\zeta)$ to be nonempty

In this section, X is a closed domain (=the closure of a connected open set) in $\mathbb{R}^n$ and ζ is a bounded continuous function on $X \times X$ vanishing on the diagonal:

$$\zeta(x, x) = 0, \quad x = (x_1, \ldots, x_n) \in X. \tag{3.1}$$

Our goal is to find conditions for

$$Q_0(\zeta) = \{u \in \mathbb{R}^X : u(x) - u(z) \leq \zeta(x, z) \quad \forall x, z \in X\}$$

to be nonempty. In case of open X, this question was studied earlier in Levin (1990b).

Theorem 3.1. *Suppose that ζ satisfies (3.1) and is continuously differentiable on some closed neighborhood $\overline{G}$ of the set $D_0 := D \cap \operatorname{int}(X \times X) = \{(x, x) : x \in \operatorname{int} X\}$. (Here $G \subset \operatorname{int}(X \times X)$ is an open neighborhood of D_0, and D stands for the diagonal $\{(x, x) : x \in X\}$ in $X \times X$; therefore $D \subset \overline{G}$, and for $(x, y) \in G$ there exist partial derivatives $\partial\zeta(x, z)/\partial x_i$, $\partial\zeta(x, z)/\partial z_i$, $i = 1, \ldots, n$, which are continuous on G and can be (uniquely) extended with preserving continuity to $\overline{G}$.) Then either $Q_0(\zeta)$ is empty or there exists a (continuously differentiable) function $u(x)$, unique up to a constant term, that satisfies the equation*

$$\nabla u(x) = \nabla_x \zeta(x, z)|_{z=x}. \tag{3.2}$$

In the latter case,

$$Q_0(\zeta) = \{u(\cdot) + \alpha : \alpha \in \mathbb{R}^1\}. \tag{3.3}$$

Proof. Suppose that $Q_0(\zeta)$ is nonempty and consider any $u \in Q_0(\zeta)$. We have $-\zeta(z, x) \leq u(x) - u(z) \leq \zeta(x, z)$ for all $x, z \in X$, which along with (3.1) implies that u is continuous on X. Fix now a point $x \in \operatorname{int} X$ and a vector $p \in \mathbb{R}^n$. Then for small $t > 0$ one has $x \pm tp \in X$, $(x, x \pm tp), (x \pm tp, x) \in G$, and

$$-\frac{\zeta(x, x + tp) - \zeta(x, x)}{t} \leq \frac{u(x + tp) - u(x)}{t} \leq \frac{\zeta(x + tp, x) - \zeta(x, x)}{t}.$$

Consequently,

$$\begin{aligned} -\nabla_z \zeta(x, z)|_{z=x} \cdot p &\leq \liminf_{t \downarrow 0} \frac{u(x + tp) - u(x)}{t} \\ &\leq \limsup_{t \downarrow 0} \frac{u(x + tp) - u(x)}{t} \leq \nabla_x \zeta(x, z)|_{z=x} \cdot p. \end{aligned} \tag{3.4}$$

Further, differentiating (3.1) yields

$$\frac{\partial\zeta}{\partial x_i}(x, z)|_{z=x} + \frac{\partial\zeta}{\partial z_i}(x, z)|_{z=x} = 0, \quad i = 1, \ldots, n,$$

i.e.,

$$\nabla_x \zeta(x, z)|_{z=x} = -\nabla_z \zeta(x, z)|_{z=x}. \tag{3.5}$$

It follows from (3.4) and (3.5) that

$$\lim_{t \downarrow 0} \frac{u(x+tp)-u(x)}{t} = \nabla_x \zeta(x,z)|_{z=x} \cdot p,$$

and since

$$\lim_{t \downarrow 0} \frac{u(x-tp)-u(x)}{-t} = -\lim_{t \downarrow 0} \frac{u(x+t(-p))-u(x)}{t}$$
$$= -\nabla_x \zeta(x,z)|_{z=x} \cdot (-p) = \nabla_x \zeta(x,z)|_{z=x} \cdot p,$$

we obtain the existence of the derivative

$$\left.\frac{du(x+tp)}{dt}\right|_{t=0} = \nabla_x \zeta(x,z)|_{z=x} \cdot p.$$

Then u is continuously differentiable on int$X$, and its gradient is given by (3.2). Therefore, this gradient is extended with preserving (3.2) up to the boundary of X hence it can be considered as a continuous vector function on X. The existence of a continuously differentiable function having the given gradient implies that such a function is uniquely determined up to a constant term; therefore, (3.3) holds true. □

Fix an arbitrary function u on X and consider for each $x \in X$ the function g^x on X,

$$g^x(z) := u(z) + \zeta(x,z), \quad z \in X. \tag{3.6}$$

The following equivalence results immediately from (3.1) and the definition of $Q_0(\zeta)$:

$$u \in Q_0(\zeta) \Leftrightarrow g^x(x) = \min\{g^x(z) : z \in X\}. \tag{3.7}$$

By analogy with Levin (1990b, 1997a, 1996) where the case of open domain X is considered, we say that a function ζ is *regular* on the closure $\overline{G}$ of an open neighborhood $G \subset \text{int}(X \times X)$ of the set $D_0 = \{(x,x) : x \in \text{int}X\} = D \cap \text{int}(X \times X)$ if:

(a) ζ is continuously differentiable on $\overline{G}$;

(b) on D there exist continuous partial derivatives $\partial^2\zeta/\partial x_i \partial x_j, \partial^2\zeta/\partial x_i \partial z_j$ and

$$\frac{\partial^2 \zeta}{\partial x_i \partial x_j}(x,x) = \frac{\partial^2 \zeta}{\partial x_j \partial x_i}(x,x), \quad i,j = 1,\ldots,n; \tag{3.8}$$

(c) for every $x \in X$, the function $\zeta(x,\cdot)$ is twice continuously differentiable on $\overline{G}(x) := \{z : (x,z) \in \overline{G}\}$.

Recall that a function on the closure $\overline{G}$ of an open set G is called continuously differentiable (C^1) or twice continuously differentiable (C^2) if at points of G there exist the corresponding partial derivatives which are continuous on G and can be (uniquely) extended with preserving continuity to $\overline{G}$. Similarly, (b) means that in D_0 there exist continuous partial derivatives $\partial^2\zeta/\partial x_i \partial x_j$

satisfying (3.8), and $\partial^2\zeta/\partial x_i\partial z_j$, and these derivatives can be extended with preserving continuity and (3.8) to the whole of D.

All conditions (a),(b),(c) are clearly satisfied when ζ is C^2 on $\overline{G}$.

If ζ is regular on $\overline{G}$, then for $(x,z) \in \overline{G}$ the matrix $(\beta_{ij}(x,z))$ is defined as follows:

$$\beta_{ij}(x,z) = \beta^{\zeta}_{ij}(x,z) := \frac{\partial^2\zeta}{\partial x_i \partial x_j}(z,z) + \frac{\partial^2\zeta}{\partial x_i \partial z_j}(z,z) + \frac{\partial^2\zeta}{\partial z_i \partial z_j}(x,z). \quad (3.9)$$

If ζ is regular on $\overline{G}$ and a function u on X satisfies (3.2), then u is C^2 on X and

$$\frac{\partial^2 u(x)}{\partial x_i \partial x_j} = \frac{\partial^2\zeta(x,x)}{\partial x_i \partial x_j} + \frac{\partial^2\zeta(x,x)}{\partial x_i \partial z_j}. \quad (3.10)$$

Taking into account (3.10), we get

$$\frac{\partial^2 g^x(z)}{\partial z_i \partial z_j} = \beta_{ij}(x,z). \quad (3.11)$$

Theorem 3.2. (necessary conditions). *Suppose that X is a closed domain in $\mathbb{R}^n$ and that the cost function ζ satisfies (3.1) and is regular on the closure $\overline{G}$ of some open neighborhood $G \subset \operatorname{int}(X \times X)$ of D_0. If $Q_0(\zeta)$ is nonempty, then for every $x \in X$, the equalities*

$$\frac{\partial^2\zeta}{\partial x_i \partial z_j}(x,x) = \frac{\partial^2\zeta}{\partial x_j \partial z_i}(x,x) \quad \forall i,j \in \{1,\dots,n\} \quad (3.12)$$

hold and the matrix $(\beta^{\zeta}_{ij}(x,x))$ is (symmetric and) positive semidefinite.

Remark 3.1. If ζ is regular on $\overline{G}$, then

$$\beta^{\zeta}_{ij}(x,x) = -\frac{\partial^2\zeta}{\partial z_i \partial x_j}(x,x). \quad (3.13)$$

This is obtained by repeated differentiation of (3.1).

Proof. Fix $u \in Q_0(\zeta)$ and consider for every $x \in \operatorname{int}X$ the function g^x on $\overline{G}(x)$, as defined by (3.6). It follows from (3.7) that

$$g^x(x) = \min\{g^x(z) : z \in \overline{G}(x)\}. \quad (3.14)$$

Furthermore, according to Theorem 3.1, the function u satisfies (3.2). This, along with the regularity assumptions on ζ, implies that u is C^2 on X, hence

$$\frac{\partial^2 u}{\partial x_i \partial x_j} = \frac{\partial^2 u}{\partial x_j \partial x_i}, \quad (3.15)$$

and (3.10) holds true. Therefore, (3.15) can be rewritten as (3.12). Furthermore, g^x is C^2 on $\overline{G}(x)$, and from (3.14) it follows that

$$\nabla g^x(z)|_{z=x} = 0 \tag{3.16}$$

(this is a condition for $z = x$ to be a stationary point of g^x; the condition coincides with (3.2)), and that the matrix $(\partial^2 g^x(x)/\partial z_i \partial z_j)$ is positive semidefinite (this is a second-order condition for $z = x$ to be the minimum point of g^x). Taking into account (3.11), we see that, in case of $x \in \text{int}X$, the matrix $(\beta_{ij}(x,x))$ is positive semidefinite, and the theorem is thus proved. In general case, we take a sequence $x^k \in \text{int}X$, $x^k \to x$. Then for any i, j one has $\partial^2\zeta(x^k, x^k)/\partial x_i \partial z_j \to \partial^2\zeta(x,x)/\partial x_i \partial z_j$, which implies (3.12), and $\beta_{ij}(x^k, x^k) \to \beta_{ij}(x,x)$, which implies the positive semidefiniteness condition:

$$\sum_{i=1}^{n}\sum_{j=1}^{n} \beta_{ij}(x,x)\xi_i\xi_j = \lim_{k\to\infty} \sum_{i=1}^{n}\sum_{j=1}^{n} \beta_{ij}(x^k,x^k)\xi_i\xi_j \geq 0 \quad \forall(\xi_1,\ldots,\xi_n).$$

The proof is complete. □

In the next theorem we consider a simply connected closed domain X and give a criterion for $Q_0(\zeta)$ to be nonempty.

Theorem 3.3. (necessary and sufficient conditions). *Suppose that X is a simply connected closed domain in $\mathbb{R}^n$ and that the cost function ζ satisfies (3.1) and is regular on the closure $\overline{G}$ of some open neighborhood $G \subset \text{int}(X \times X)$ of $D_0 = \{(x,x) : x \in \text{int}X\} = D \cap \text{int}(X \times X)$. Then $Q_0(\zeta)$ is nonempty if and only if:*

(a) equalities (3.12) hold true, and

(b) for any pair of points $x, z \in X$ and for any piecewise smooth oriented curve $\gamma(z,x)$ lying in X and leading from z to x, the inequality

$$\int_{\gamma(z,x)} \nabla_\xi \zeta(\xi,\eta)|_{\eta=\xi} \cdot d\xi \leq \zeta(x,z) \tag{3.17}$$

holds true.

Remark 3.2. From (3.12) and the regularity assumption the equality $\nabla_x\zeta(x,z)|_{z=x} = \nabla u(x)$ results, where u is a smooth (C^2) function on X, and as X is assumed to be simply connected, the integral in (3.17) depends on points x and z only and does not depend on the curve $\gamma(z,x)$. Therefore, for specific domains one can choose the simplest curves. For example, if X is convex, then $\gamma(z,x)$ can be taken to be the segment zx so that $\gamma(z,x) = \{z + t(x-z) : 0 \leq t \leq 1\}$. In such a case, $d\zeta = (x-z)dt$ and (3.17) is rewritten as follows:

$$\int_0^1 \sum_{i=1}^n \frac{\partial\zeta(z+t(x-z), z+t(x-z))}{\partial x_i}(x_i - z_i)\,dt \le \zeta(x,z). \tag{3.18}$$

Similarly, if X is star-like with respect to some point $x_0 \in X$, then it is possible to take the two-link curve $\gamma(z,x)$ composed of the segments zx_0 and x_0x. We omit a specification of (3.17) in that case.

Proof. Necessity. If $u \in Q_0(\zeta)$, then, by Theorem 3.2, equalities (3.12) hold true. Further, from Theorem 3.1 and the regularity of ζ on $\overline{G}$ it follows that u satisfies (3.2) and is C^2 on X. Therefore,

$$\int_{\gamma(z,x)} \nabla u(\xi)\cdot d\xi = \int_{\gamma(z,x)} \nabla_\xi \zeta(\xi,\eta)|_{\eta=\xi}\cdot d\xi = u(x) - u(z),$$

and as $u \in Q_0(\zeta)$, we see that (3.17) holds true.

Sufficiency. We consider on X the vector field

$$\nabla_x\zeta(x,z)|_{z=x} = \left(\frac{\partial\zeta}{\partial x_1}(x,x),\dots,\frac{\partial\zeta}{\partial x_n}(x,x)\right).$$

Using the regularity of ζ and (3.12), we get that for any $i,j \in \{1,\dots,n\}$

$$\frac{\partial}{\partial x_j}\left(\frac{\partial\zeta}{\partial x_i}(x,x)\right) = \frac{\partial}{\partial x_i}\left(\frac{\partial\zeta}{\partial x_j}(x,x)\right).$$

This and the simple connectedness of X give us that for any $x_0 \in X$ the function

$$u(x) := \int_{\gamma(x_0,x)} \nabla_\xi\zeta(\xi,\eta)|_{\eta=\xi}\cdot d\xi$$

satisfies (3.2), and the integral

$$\int_\gamma \nabla_\xi\zeta(\xi,\eta)|_{\eta=\xi}\cdot d\xi$$

is equal to zero along any closed contour γ. Then

$$u(x) - u(z) = \int_{\gamma(z,x)} \nabla_\xi\zeta(\xi,\eta)|_{\eta=\xi}\cdot d\xi,$$

and (3.17) implies $u \in Q_0(\zeta)$. □

In Theorems 3.4 and 3.5 below we consider a closed convex domain X and present sufficient conditions for $Q_0(\zeta)$ to be nonempty.

Theorem 3.4. (sufficient conditions). *Let X be a closed convex domain. Suppose that ζ satisfies (3.1) and is regular on $X \times X$. Suppose also that equalities (3.12) hold true and that for every $x, z \in X$ the matrix $(\beta_{ij}^\zeta(x,z))$ is positive semidefinite. Then $Q_0(\zeta)$ is nonempty.*

Proof. From (3.12) and the regularity of ζ on $X \times X$ it follows the existence of a function u on X satisfying (3.2). Let us construct functions g^x, $x \in X$, using that u. In view of (3.11), positive semidefiniteness of the matrix $(\beta^{\zeta}_{ij}(x, z))$ means that every g^x is convex. It follows from (3.2) that any $x \in \text{int}X$ is a stationary point of g^x, and since g^x is convex, x is a minimum point of g^x. Then, by continuity, every $x \in X$ is a minimum point of g^x, and the proof is concluded by applying the equivalence (3.7). □

Theorem 3.5. (sufficient conditions). *Let X be a closed convex domain. Suppose that ζ satisfies (3.1) and is regular on the closure $\overline{G}$ of some open neighborhood $G \subset \text{int}(X \times X)$ of $D_0 = \{(x, x) : x \in \text{int}X\}$. Suppose also that equalities (3.12) hold true and that for every $x, z \in X$*

$$(\nabla_z\zeta(x, z) - \nabla_z\zeta(z, z)) \cdot (z - x) = \sum_{i=1}^{n} \left(\frac{\partial\zeta(x, z)}{\partial z_i} - \frac{\partial\zeta(z, z)}{\partial z_i} \right) (z_i - x_i) \geq 0. \tag{3.19}$$

Then $Q_0(\zeta)$ is nonempty.

Proof. Equality (3.1) implies (3.5), while (3.12) and the simple connectedness of X imply the existence of a function u satisfying (3.2). We fix an arbitrary point $z \in X$ and let

$$z_t := (1 - t)x + tz, \quad 0 \leq t \leq 1.$$

Consider the function $g^x(z) = u(z) + \zeta(x, z)$. Using (3.2) and (3.5), we get that

$$\begin{aligned} \frac{dg^x(z_t)}{dt} &= \nabla g^x(z_t) \cdot \frac{dz_t}{dt} = (\nabla_z\zeta(x, z_t) - \nabla_z\zeta(z_t, z_t)) \cdot (z - x) \\ &= \frac{1}{t}(\nabla_z\zeta(x, z_t) - \nabla_z\zeta(z_t, z_t)) \cdot (z_t - x). \end{aligned}$$

In view of (3.19), $dg^x(z_t)/dt \geq 0$; consequently,

$$g^x(z) = g^x(x) + \int_0^1 \frac{dg^x(z_t)}{dt} \, dt \geq g^x(x),$$

and, by virtue of (3.7), $u \in Q_0(\zeta)$. □

Remark 3.3. Let us show that Theorem 3.4 follows from Theorem 3.5. From (3.12) and regularity of ζ it follows the existence of a smooth function u that satisfies (3.2). Let us consider functions $t \mapsto g^x(z_t) = u(z_t) + \zeta(x, z_t)$, where $z_t = (1 - t)x + tz$, $t \in [0, 1]$. Then (3.2) can be rewritten as

$$\frac{dg^x(z_0)}{dt} := \lim_{t \downarrow 0} \frac{g^x(z_t) - g^x(z_0)}{t} = 0. \tag{3.20}$$

Taking into account (3.11), we get

$$\begin{aligned}\frac{d^2 g^x(z_t)}{dt^2} &= \frac{d}{dt}\left(\nabla g^x(z_t)\cdot\frac{dz_t}{dt}\right) = \frac{d}{dt}\nabla g^x(z_t)\cdot(z-x)\\ &= \frac{d}{dt}\sum_{i=1}^n \frac{\partial g^x(z_t)}{\partial z_i}(z_i - x_i)\\ &= \sum_{i=1}^n\sum_{j=1}^n \frac{\partial^2 g^x(z_t)}{\partial z_i \partial z_j}(z_j - x_j)(z_i - x_i)\\ &= \sum_{i=1}^n\sum_{j=1}^n \beta_{ij}(x, z_t)(z_i - x_i)(z_j - x_j);\end{aligned}$$

therefore the hypothesis that the matrix $(\beta_{ij}^{\zeta}(x,z))$ is positive semidefinite means convexity of the function $t \mapsto g^x(z_t)$ on $[0,1]$. That convexity along with (3.20) implies

$$\frac{dg^x(z_t)}{dt} \geq 0 \quad \forall t, 0 < t < 1.$$

Now, as

$$\frac{dg^x(z_t)}{dt} = \frac{1}{t}(\nabla_z \zeta(x, z_t) - \nabla_z \zeta(z_t, z_t))\cdot(z_t - x)$$

(see proof of Theorem 3.5), we obtain (3.19) by tending t to 1. The hypotheses of Theorem 3.4 are thus stronger than those of Theorem 3.5.

4. Characterizing optimal Monge solutions to MKPs with smooth cost functions

We return now to the Monge–Kantorovich problem $MKP(c;\sigma_1,\sigma_2)$. Recall that $X \subset \mathbb{R}^n$ and $Y \subset \mathbb{R}^m$ are closed domains with positive Borel measures σ_1 on X and σ_2 on Y, $\sigma_1 X = \sigma_2 Y$. In this section, c is assumed to be a bounded smooth (C^2) real–valued function on $X \times Y$, and $f : X \to Y$ is assumed to be a smooth (C^1) measure–preserving map. We shall consider the function $\varphi : X \times X \to \mathbb{R}$,

$$\varphi(x,z) = c(z, f(x)) - c(x, f(x)) \tag{4.1}$$

(cf. (2.3)). Given $x = (x_1, \ldots, x_n)$ and $z = (z_1, \ldots, z_n)$ in X, we have

$$f(x) = (f_1(x_1,\ldots,x_n),\ldots,f_m(x_1,\ldots,x_n)),$$

$$\begin{aligned}\varphi(x,z) &= \varphi(x_1,\dots,x_n,z_1,\dots,z_n)\\ &= c(z_1,\dots,z_n,f_1(x_1,\dots,x_n),\dots,f_m(x_1,\dots,x_n))\\ &\quad - c(x_1,\dots,x_n,f_1(x_1,\dots,x_n),\dots,f_m(x_1,\dots,x_n)).\end{aligned}$$

It follows from (4.1) that φ is regular on $X \times X$ (for the definition of regularity, see Section 3). Moreover, if f is C^2 then the function $\psi : X \times X \to \mathbb{R}$,

$$\psi(x,z) := \varphi(z,x) = c(x,f(z)) - c(z,f(z)), \tag{4.2}$$

is regular on $X \times X$ as well. Clearly, both functions, φ and ψ, vanish on the diagonal, and

$$Q_0(\psi) = -Q_0(\varphi). \tag{4.3}$$

Then

$$Q_0(\psi|_{Z\times Z}) = -Q_0(\varphi|_{Z\times Z}), \tag{4.4}$$

where $Z = \operatorname{spt}(\sigma_1)$.

For every $(x,z) = (x_1,\dots,x_n,z_1,\dots,z_n) \in X \times X$ we consider the $n \times n$ matrices $(\beta^{\zeta}_{ij}(x,z))$,

$$\beta^{\zeta}_{ij}(x,z) = \frac{\partial^2\zeta(z,z)}{\partial x_i \partial x_j} + \frac{\partial^2\zeta(z,z)}{\partial x_i \partial z_j} + \frac{\partial^2\zeta(x,z)}{\partial z_i \partial z_j}, \quad \zeta = \varphi,\psi \tag{4.5}$$

(cf. (3.9)). A direct computation yields

$$\beta^{\varphi}_{ij}(x,z) = \frac{\partial^2 c(z,f(x))}{\partial x_i \partial x_j} - \frac{\partial^2 c(z,f(z))}{\partial x_i \partial x_j} - \sum_{k=1}^{m} \frac{\partial^2 c(z,f(z))}{\partial x_i \partial y_k}\frac{\partial f_k(z)}{\partial x_j}, \tag{4.6}$$

and (for f being C^2)

$$\begin{aligned}\beta^{\psi}_{ij}(x,z) = &\sum_{k=1}^{m}\sum_{l=1}^{m}\left(\frac{\partial^2 c(x,f(z))}{\partial y_k \partial y_l} - \frac{\partial^2 c(z,f(z))}{\partial y_k \partial y_l}\right)\frac{\partial f_k(z)}{\partial x_i}\frac{\partial f_l(z)}{\partial x_j}\\ &+\sum_{k=1}^{m}\left(\frac{\partial c(x,f(z))}{\partial y_k} - \frac{\partial c(z,f(z))}{\partial y_k}\right)\frac{\partial^2 f_k(z)}{\partial x_i \partial x_j}\\ &-\sum_{k=1}^{m}\frac{\partial^2 c(z,f(z))}{\partial x_j \partial y_k}\frac{\partial f_k(z)}{\partial x_i}.\end{aligned} \tag{4.7}$$

It follows that

$$\beta^{\varphi}_{ij}(x,x) = \beta^{\psi}_{ji}(x,x) = -\sum_{k=1}^{m}\frac{\partial^2 c(x,f(x))}{\partial x_i \partial y_k}\frac{\partial f_k(x)}{\partial x_j}. \tag{4.8}$$

The following optimality conditions for Monge solutions of MKP are derived from Corollary 2.2 and nonemptyness conditions for $Q_0(\zeta)$ as given in Section 3.

Theorem 4.1. (necessary conditions). *Suppose that c is bounded and C^2 on $X \times Y$,* spt$(\sigma_1) = X$*, f is C^1, and $f(\sigma_1) = \sigma_2$. If $\mu_f = \mu_f(\sigma_1)$ is an optimal Monge solution to MKP, then for every $x \in X$ the matrix*

$$\left(\sum_{k=1}^{m} \frac{\partial^2 c(x, f(x))}{\partial x_i \partial y_k} \frac{\partial f_k(x)}{\partial x_j}\right)_{ij} \tag{4.9}$$

is symmetric and negative semidefinite. In such a case, there exists a function $u \in Q_0(\varphi) = -Q_0(\psi)$, which is unique up to an additive constant. The function is C^2 on X and it is connected with c and f by the equality

$$\nabla u(x) = -\nabla_x c(x, y)|_{y=f(x)}. \tag{4.10}$$

Proof. By Corollary 2.2, there is a function $u \in Q_0(\varphi)$. Then, by Theorem 3.1, equality (3.2) holds with $\zeta = \varphi$, which is rewritten as (4.10). Furthermore, it follows from Theorem 3.2 that, for every $x \in X$, the matrix $(\beta_{ij}^{\varphi}(x, x))_{ij}$ is symmetric and positive semidefinite. Now it remains to notice that, in view of (4.6), this matrix is precisely the matrix (4.9) taken with the sign minus. □

Remark 4.1. In Theorem 4.1, the requirement that c is bounded on $X \times Y$ can be weakened to the assumption that the functions $\varphi(\cdot, z), \psi(\cdot, z)$ are σ_1-summable whenever $z \in \pi_1$spt$(\mu_f) = \{z' : (z', f(z')) \in$ spt$(\mu_f)\}$.

Theorem 4.2. (necessary and sufficient conditions). *Suppose that a closed domain X is simply connected, c is bounded and C^2 on $X \times Y$,* spt$(\sigma_1) = X$*, f is C^1, and $f(\sigma_1) = \sigma_2$. The following statements are equivalent:*

(a) $\mu_f = \mu_f(\sigma_1)$ is an optimal Monge solution to MKP;

(b) for every $x \in X$, the matrix (4.9) is symmetric, and for every $x, z \in X$ and any piecewise smooth oriented curve $\gamma(z, x)$ lying in X and leading from z to x, the inequality

$$\int_{\gamma(z,x)} \nabla_\xi c(\xi, f(\xi)) \cdot d\xi \geq c(x, f(x)) - c(z, f(x)) \tag{4.11}$$

holds true;

(c) for every $x \in X$, the matrix (4.9) is symmetric, and for every $x, z \in X$ and any piecewise smooth oriented curve $\gamma(z, x)$ lying in X and leading from z to x, the inequality

$$\int_{\gamma(z,x)} \nabla_\xi c(\xi, f(\xi)) \cdot d\xi \leq c(x, f(z)) - c(z, f(z)) \tag{4.12}$$

holds true.

Proof. The equivalence (a)⇔(b) follows from Corollary 2.2 and Theorem 3.3 for $\zeta = \varphi$ if one takes into account that

$$\frac{\partial^2 \varphi(x,x)}{\partial x_i \partial z_j} = \sum_{k=1}^{m} \frac{\partial^2 c(x, f(x))}{\partial x_j \partial y_k} \frac{\partial f_k(x)}{\partial x_i} \tag{4.13}$$

and that

$$\frac{\partial \varphi(x,x)}{\partial x_i} = -\frac{\partial c(x, f(x))}{\partial x_i},$$

hence

$$\nabla_x \varphi(x,z)|_{z=x} = -\nabla_x c(x, f(z))|_{z=x}.$$

Similarly, the equivalence (a)⇔(c) is established by the same arguments applied to $\zeta = \psi$ if one takes into account (4.3) and the equality

$$\nabla_x \psi(x,z)|_{z=x} = \nabla_x c(x, f(z))|_{z=x},$$

which follows from (4.2). □

Theorem 4.3. (sufficient conditions). *Suppose that a closed domain X is convex, c is bounded and C^2 on $X \times Y$, f is C^1, and $f(\sigma_1) = \sigma_2$. The measure $\mu_f(\sigma_1)$ is an optimal solution to MKP in any of two cases:*

(a) the matrices (4.9) are symmetric and the inequality

$$\sum_{j=1}^{n} \left(\frac{\partial c(z, f(z))}{\partial x_j} - \frac{\partial c(z, f(x))}{\partial x_j} \right) (z_j - x_j) \le 0; \tag{4.14}$$

holds for any $x, z \in X$;

(b) the matrices (4.9) are symmetric and the inequality

$$\sum_{j=1}^{n} \sum_{k=1}^{m} \left(\frac{\partial c(z, f(z))}{\partial y_k} - \frac{\partial c(x, f(z))}{\partial y_k} \right) \frac{\partial f_k(z)}{\partial x_j} (z_j - x_j) \le 0. \tag{4.15}$$

holds for any $x, z \in X$.

Proof. (a) Taking into account (4.12) we see that equalities (3.12) hold true for $\zeta = \varphi$. Furthermore, since

$$\nabla_z \varphi(x,z) = \nabla_x c(z, f(x))$$

(recall that $\nabla_x c(z, f(x)) := \nabla_x c(x,y)|_{x=z, y=f(x)}$), (4.13) can be rewritten as (3.19) for $\zeta = \varphi$. By Theorem 3.5, $Q_0(\varphi)$ is then nonempty, and the optimality of μ_f results from Corollary 2.2.

(b) Similarly, equalities (3.12) hold true for $\zeta = \psi$, and since, in view of (4.2),

$$\frac{\partial\psi(x,z)}{\partial z_j} = \sum_{k=1}^{m}\left(\frac{\partial c(x,f(z))}{\partial y_k} - \frac{\partial c(z,f(z))}{\partial y_k}\right)\frac{\partial f_k(z)}{\partial x_j} - \frac{\partial c(z,f(z))}{\partial x_j},$$

we get

$$\sum_{j=1}^{n}\left(\frac{\partial\psi(x,z)}{\partial z_j} - \frac{\partial\psi(z,z)}{\partial z_j}\right)(z_j - x_j)$$
$$= \sum_{j=1}^{n}\sum_{k=1}^{m}\left(\frac{\partial c(x,f(z))}{\partial y_k} - \frac{\partial c(z,f(z))}{\partial y_k}\right)\frac{\partial f_k(z)}{\partial x_j}(z_j - x_j);$$

therefore, (4.14) can be rewritten as (3.19) for $\zeta = \psi$. From Theorem 3.5 it follows that $Q_0(\psi)$ is nonempty, and applying Corollary 2.2 along with (4.3) concludes the proof. □

More strong sufficient conditions are presented in the next theorem.

Theorem 4.4. (sufficient conditions). *Suppose that a closed domain X is convex, c is bounded and C^2 on $X \times Y$, f is C^1, and $f(\sigma_1) = \sigma_2$. The measure $\mu_f(\sigma_1)$ is an optimal solution to MKP in any of two cases:*

(a) the matrices (4.9) are symmetric and the matrices $(\beta^{\varphi}_{ij}(x,z))$, $x, z \in X$, are positive semidefinite;

(b) f is C^2, the matrices (4.9) are symmetric, and the matrices $(\beta^{\psi}_{ij}(x,z))$, $x, z \in X$, are positive semidefinite.

Proof. Using Theorem 3.4 with $\zeta = \varphi$ in case (a) or with $\zeta = \psi$ in case (b) shows that $Q_0(\zeta)$ is nonempty, and applying Corollary 2.2 (along with (4.3) in case (b)) concludes the proof. □

5. Explicit solutions to Monge problems: theory

In this section, X and Y are supposed to be bounded closed domains in $\mathbb{R}^n$ and $\mathbb{R}^m$, respectively. Recall that if μ_f is an optimal solution to $MKP(c;\sigma_1,\sigma_2)$, then f is an optimal solution to $MP(c;\sigma_1,\sigma_2)$.

Theorem 5.1. *Suppose that $X \subset \mathbb{R}^n$ and $Y \subset \mathbb{R}^n$ are bounded closed domains and that a function $c(x,y) = \sum_1^n x_j b_j(y)$ and a map $f \in \Phi(\sigma_1,\sigma_2)$ are smooth (C^2 and C^1, respectively). Then:*

I. If $\mathrm{spt}(\sigma_1) = X$ *and $\mu_f(\sigma_1)$ is an optimal solution to $MKP(c;\sigma_1,\sigma_2)$, then the matrices*

$$\left(\sum_{k=1}^{n}\frac{\partial b_i(f(x))}{\partial y_k}\frac{\partial f_k(x)}{\partial x_j}\right)_{ij}, \quad x \in X, \tag{5.1}$$

are symmetric and negative semidefinite;

II. If X is convex and matrices (5.1) are symmetric and negative semidefinite, then $\mu_f(\sigma_1)$ is an optimal solution to $MKP(c;\sigma_1,\sigma_2)$.

Proof. I. This follows from Theorem 4.1 if one takes into account that for a given function c the matrix (4.9) is rewritten as (5.1).

II. We derive from (4.6) that, for a given c,

$$\begin{aligned}\beta_{ij}^{\varphi}(x,z) = \beta_{ij}^{\varphi}(z,z) &= -\sum_{k=1}^{m} \frac{\partial^2 c(x,f(x))}{\partial x_i \partial y_k} \frac{\partial f_k(x)}{\partial x_j} \\ &= -\sum_{k=1}^{n} \frac{\partial b_i(f(x))}{\partial y_k} \frac{\partial f_k(x)}{\partial x_j},\end{aligned}$$

and the optimality of μ_f follows then from Theorem 4.4, (a). □

If $b_i(y) = \pm y_i$ then (5.1) becomes $(\pm \partial f_i(x)/\partial x_j)$, and we obtain the following corollary.

Corollary 5.1. *Suppose that X is a convex bounded closed domain, Y is a bounded closed domain, $c(x,y) = \sum_1^n x_j y_j$ or $c(x,y) = -\sum_1^n x_j y_j$, $f : X \to Y$ is C^1,* spt$(\sigma_1) = X$. *A measure $\mu_f(\sigma_1)$ is an optimal solution to $MKP(c;\sigma_1,f(\sigma_1))$ if and only if there exists a smooth concave function $u(x)$ on X such that $f(x) = \nabla u(x)$ in case of $c(x,y) = \sum_1^n x_j y_j$ or $f(x) = -\nabla u(x)$ in case of $c(x,y) = -\sum_1^n x_j y_j$. In both cases, $-u \in Q_0(\varphi)$ hence $u \in Q_0(\psi)$, where φ and ψ are defined by (4.1) and (4.2), respectively. Moreover, if σ_1 is absolutely continuous with respect to the Lebesgue measure on X, then $\mu_f(\sigma_1)$ is the unique optimal solution to $MKP(c;\sigma_1,f(\sigma_1))$ and f is the unique, up to values on a σ_1-negligible set, optimal solution to the corresponding Monge problem $MP(c;\sigma_1,f(\sigma_1))$.*

Here, the uniqueness of optimal solutions in case of absolutely continuous σ_1 follows from Theorem 1.4.

It is easily seen that if $\mu_f(\sigma_1)$ is an optimal solution to $MKP(c;\sigma_1,f(\sigma_1))$, then it remains to be an optimal solution to $MKP(c';\sigma_1,f(\sigma_1))$, where $c'(x,y) = c(x,y) + a(x) + b(y)$. Hence the next result follows.

Corollary 5.2. *The assertions of Corollary 5.1 for the case of $c(x,y) = -\sum_1^n x_j y_j$ hold true for the cost function $c(x,y) = (x-y)\cdot(x-y) = \sum_1^n (x_j - y_j)^2$. Similarly, the assertions of Corollary 5.1 for the case of $c(x,y) = \sum_1^n x_j y_j$ hold true for the cost function $c(x,y) = -(x-y)\cdot(x-y)$.*

Corollary 5.3. *I. Suppose that $c(x,y) = (x-y)\cdot(x-y) = \sum_1^n (x_j - y_j)^2$, and let $f(x) = Ax + b$ where A is a linear operator $\mathbb{R}^n \to \mathbb{R}^n$, $b \in \mathbb{R}^n$. The measure $\mu_f(\sigma_1)$ is an optimal solution to $MKP(c;\sigma_1,f(\sigma_1))$ if and* *only if A is symmetric and positive semidefinite. Moreover, if σ_1 is absolutely continuous with respect to the Lebesgue measure on X and A is symmetric positive semidefinite, then $\mu_f(\sigma_1)$ is the unique optimal solution to $MKP(c; \sigma_1, f(\sigma_1))$ and f is the unique, up to values on a σ_1-negligible set, optimal solution to the corresponding Monge problem $MP(c; \sigma_1, f(\sigma_1))$.*

II. Similarly, suppose that $c(x,y) = -(x-y)\cdot(x-y) = -\sum_1^n (x_j - y_j)^2$, and let $f(x) = Ax + b$ where A is a linear operator $\mathbb{R}^n \to \mathbb{R}^n$, $b \in \mathbb{R}^n$. The measure $\mu_f(\sigma_1)$ is an optimal solution to $MKP(c; \sigma_1, f(\sigma_1))$ if and only if A is symmetric and negative semidefinite. Moreover, if σ_1 is absolutely continuous with respect to the Lebesgue measure on X and A is symmetric negative semidefinite, then $\mu_f(\sigma_1)$ is the unique optimal solution to $MKP(c; \sigma_1, f(\sigma_1))$ and f is the unique, up to values on a σ_1-negligible set, optimal solution to the corresponding Monge problem $MP(c; \sigma_1, f(\sigma_1))$.

Proof. I. According to Corollary 5.2, optimality of f is equivalent to the existence of a smooth convex function $u(x)$ on X such that $Ax + b = \nabla u(x)$, and it remains to notice that such a function exists if and only if A is symmetric positive semidefinite, in what case $u(x) = \frac{1}{2}Ax \cdot x + b \cdot x + \alpha$, $\alpha \in \mathbb{R}$.

II. Similarly, again by Corollary 5.2, optimality of f is equivalent to the existence of a smooth concave function $u(x)$ on X such that $Ax + b = \nabla u(x)$, and it remains to notice that such a function exists if and only if A is symmetric negative semidefinite, in what case $u(x) = \frac{1}{2}Ax \cdot x + b \cdot x + \alpha$, $\alpha \in \mathbb{R}$. □

Let us consider now general cost functions having the form $c(x,y) = h(x - y)$, where X, Y are bounded closed domains in $\mathbb{R}^n$ and h is a smooth (C^2) function on

$$X - Y := \{z = x - y : x \in X, y \in Y\}.$$

Theorem 5.2. *Suppose that X, Y are bounded closed domains in $\mathbb{R}^n$, h is a smooth (C^2) function on $X - Y$, $c(x,y) = h(x-y)$, and $f \in \Phi(\sigma_1, \sigma_2)$ is a smooth (C^1) map. The following statements hold true:*

I. If $\mu_f(\sigma_1)$ is an optimal solution to $MKP(c; \sigma_1, \sigma_2)$ and $\mathrm{spt}(\sigma_1) = X$, then the matrices

$$\left(\sum_{k=1}^n \frac{\partial^2 h(x - f(x))}{\partial x_i \partial x_k} \frac{\partial f_k(x)}{\partial x_j} \right)_{ij}, \quad x \in X \tag{5.2}$$

are symmetric and positive semidefinite.

II. If X is convex, matrices (5.2) are symmetric, and the inequality

$$(\nabla h(z - f(z)) - \nabla h(z - f(x))) \cdot (z - x) \le 0 \tag{5.3}$$

holds true whenever $x, z \in X$, then $\mu_f(\sigma_1)$ is an optimal solution to $MKP(c; \sigma_1, \sigma_2)$.

III. If X is convex and the matrices

$$\left(\sum_{k=1}^{n} \frac{\partial^2 h(z-f(x))}{\partial x_i \partial x_k} \frac{\partial f_k(x)}{\partial x_j}\right)_{ij}, \quad x, z \in X, \tag{5.4}$$

are positive semidefinite, then (5.3) holds true for all $x, z \in X$.

Proof. I. A direct computation yields

$$\sum_{k=1}^{m} \frac{\partial^2 c(x, f(x))}{\partial x_i \partial y_k} \frac{\partial f_k(x)}{\partial x_j} = -\sum_{k=1}^{n} \frac{\partial^2 h(x-f(x))}{\partial x_i \partial x_k} \frac{\partial f_k(x)}{\partial x_j}, \tag{5.5}$$

and the statement follows then from Theorem 4.1.

II. Taking into account (5.5), this is a particular case of Theorem 4.3 (a).

III. We let $x(t) = z + t(x-z)$ and define a function $a(t)$ on $[0,1]$ as follows:

$$\begin{aligned} a(t) :&= (\nabla h(z-f(z)) - \nabla h(z - f(x(t)))\cdot(x(t)-z) \\ &= \sum_{i=1}^{n}\left(\frac{\partial h(z-f(z))}{\partial x_i} - \frac{\partial h(z-f(x(t)))}{\partial x_i}\right)(x_i(t)-z_i). \end{aligned}$$

Clearly, a is a smooth function, and as $a(1) = (\nabla h(z-f(z)) - \nabla h(z - f(x))) \cdot (x-z)$ and $a(0) = 0$, there exists θ, $0 < \theta < 1$, such that $a(1) = da(\theta)/dt$. We have $x(t) - z = t(x-z)$, hence

$$a(t) = tb(t),$$

where

$$\begin{aligned} b(t) &= (\nabla h(z-f(z)) - \nabla h(z - f(z+t(x-z)))) \cdot (x-z) \\ &= \sum_{i=1}^{n}\left(\frac{\partial h(z-f(z))}{\partial x_i} - \frac{\partial h(z-f(z+t(x-z)))}{\partial x_i}\right)(x_i - z_i). \end{aligned} \tag{5.6}$$

Therefore,

$$\begin{aligned} \frac{da(t)}{dt} &= b(t) + t\frac{db(t)}{dt} \\ &= b(t) + t\sum_{i=1}^{n}\sum_{k=1}^{n}\sum_{j=1}^{n} \frac{\partial^2 h(z-f(x(t)))}{\partial x_i \partial x_k}\frac{\partial f_k(x(t))}{\partial x_j}(x_i - z_i)(x_j - z_j). \end{aligned} \tag{5.7}$$

Thus, we have

$$(\nabla h(z-f(z)) - \nabla h(z-f(x)))\cdot(z-x) = -a(1) = -\frac{da(\theta)}{dt}. \tag{5.8}$$

Furthermore, since b is smooth and $b(0)=0$, there exists θ_1, $0<\theta_1<\theta$, such that

$$\begin{aligned} b(\theta) &= \theta\frac{db(\theta_1)}{dt} \\ &= \theta\sum_{i=1}^{n}\sum_{k=1}^{n}\sum_{j=1}^{n}\frac{\partial^2 h(z-f(x(\theta_1)))}{\partial x_i \partial x_k}\frac{\partial f_k(x(\theta_1))}{\partial x_k}(x_i-z_i)(x_j-z_j). \end{aligned} \tag{5.9}$$

Now, taking into account positive semidefiniteness of matrices (5.4), inequality (5.3) results from (5.7)–(5.9). □

Corollary 5.4. *Suppose that X is a convex bounded closed domain, $Y = f(X)$, $f(x) = x+b$ $\forall x \in X$, $c(x,y) = h(x-y)$ $\forall x \in X, y \in Y$, and h is a smooth convex function on $X-Y$. Then, for every σ_1, $\mu_f(\sigma_1)$ is an optimal solution to $MKP(c;\sigma_1, f(\sigma_1))$ hence f is an optimal solution to $MP(c;\sigma_1, f(\sigma_1))$. Moreover, if h is strictly convex, σ_1 is absolutely continuous with respect to the Lebesgue measure on X, and* spt$(\sigma_1) = X$*, then $\mu_f(\sigma_1)$ is the unique optimal solution to $MKP(c;\sigma_1, f(\sigma_1))$ and f is the unique, up to values on a σ_1-negligible set, optimal solution to the corresponding Monge problem $MP(c;\sigma_1, f(\sigma_1))$.*

Proof. We have $\partial f_k(x)/\partial x_j = 1$ if $k=j$ and $=0$ if $k\neq j$; therefore, matrices (5.4) are rewritten as

$$\left(\frac{\partial^2 h(z-x-b)}{\partial x_i \partial x_j}\right)_{ij},$$

and as h is convex, these matrices are positive semidefinite. The optimality of μ_f and f follows then from Theorem 5.2. The uniqueness of optimal solutions in the case where h is strictly convex and σ_1 is absolutely continuous with respect to the Lebesgue measure on X follows from Theorem 1.2. □

Corollary 5.5. *Suppose that X is a convex bounded closed domain, $Y = f(X)$, $f(x) = -x+b$ $\forall x \in X$, $c(x,y) = h(x-y)$ $\forall x \in X, y \in Y$, and h is a smooth concave function on $X-Y$. Then, for every σ_1, $\mu_f(\sigma_1)$ is an optimal solution to $MKP(c;\sigma_1, f(\sigma_1))$ hence f is an optimal solution to $MP(c;\sigma_1, f(\sigma_1))$. Moreover, if h is strictly concave, σ_1 is absolutely continuous with respect to the Lebesgue measure on X, and* spt$(\sigma_1) = X$*, then $\mu_f(\sigma_1)$ is the unique optimal solution to $MKP(c;\sigma_1, f(\sigma_1))$ and f is the unique, up to values on a σ_1-negligible set, optimal solution to the corresponding Monge problem $MP(c;\sigma_1, f(\sigma_1))$.*

Proof. We have $\partial f_k(x)/\partial x_j = -1$ if $k = j$ and $= 0$ if $k \neq j$; therefore, matrices (5.4) are rewritten as

$$\left(-\frac{\partial^2 h(z+x-b)}{\partial x_i \partial x_j}\right)_{ij},$$

and as h is concave, these matrices are positive semidefinite. The optimality of μ_f and f follows then from Theorem 5.2. The uniqueness of optimal solutions in the case where h is strictly concave and σ_1 is absolutely continuous with respect to the Lebesgue measure on X follows from Theorem 1.3. □

Suppose now that $X \subset \mathbb{R}^n$ and $Y \subset \mathbb{R}^m$ are bounded closed domains, c is a continuous function on $X \times Y$, and σ_1 is a positive Borel measure on X. We assume that there exists finite or countable family of closed domains, $\mathcal{X} = \{X_i\}$, $X_i \subset X$, such that either $\mathcal{X}$ is finite and $X = \bigcup_i X_i$ or $\mathcal{X}$ is countable, $\bigcup_i \operatorname{int} X_i$ is dense in X, and $X \setminus \bigcup_i \operatorname{int} X_i$ is σ_1-negligible. Also we suppose that for every pair $X_i, X_j \in \mathcal{X}$ two conditions are satisfied as follows:

i) their interiors, $\operatorname{int} X_i$ and $\operatorname{int} X_j$, have no common points, and

ii) $\sigma_1(X_i \cap X_j) = 0$.

Let σ_{1i} denote the restriction of σ_1 to X_i. Given continuous maps $f^i : X_i \to Y$, we set $\sigma_{2i} = f^i(\sigma_{1i})$ and consider on Y the measure $\sigma_2 := \sum_i \sigma_{2i}$. Taking into account i) and ii), we have

$$\sigma_1 X = \sum_i \sigma_1 X_i = \sum_i \sigma_{1i} X_i = \sum_i \sigma_{2i} Y = \sigma_2 Y.$$

We consider the Monge–Kantorovich problems $MKP(c;\sigma_1,\sigma_2)$ and $MKP(c;\sigma_{1i},\sigma_{2i})$, and define the measure

$$\mu := \sum_i \mu_{f^i}(\sigma_{1i}). \tag{5.10}$$

Let us consider a map $f : X \to Y$ that coincides with f^i on $\operatorname{int} X_i$ for every i and is defined arbitrarily on $X \setminus \bigcup_i \operatorname{int} X_i$. It is clear, in view of i),ii), that $f \in \Phi(\sigma_1,\sigma_2)$ and $\mu = \mu_f$. A question arises, when μ is an optimal solution to $MKP(c;\sigma_1,\sigma_2)$? The following two theorems answer this question. The answer is given in terms of functions φ_i on $X_i \times X$,

$$\varphi_i(x,z) := c(z, f^i(x)) - c(x, f^i(x)), \quad x \in X_i, z \in X. \tag{5.11}$$

Theorem 5.3. *Suppose that* $\operatorname{spt}(\sigma_1) = X$. *If the family* $\mathcal{X} = \{X_i\}$ *is finite and* $X = \bigcup_i X_i$, *then the following statements are equivalent:*

(a) $\mu = \mu_f$ *is an optimal solution to* $MKP(c;\sigma_1,\sigma_2)$ *hence* f *is an optimal solution to* $MP(c;\sigma_1,\sigma_2)$;

(b) there are functions $u_i \in Q_0(\varphi_i|_{X_i \times X_i})$ *and real numbers* ν_i *such that*

$$\nu_i - \nu_j \le \gamma(i,j) := \min\{\varphi_i(x,z) - u_i(x) + u_j(z) : x \in X_i, z \in X_j\} \quad \forall i,j; \tag{5.12}$$

(c) there is a continuous function u on X such that $u(x) - u(z) \le \varphi_i(x,z)$ whenever $x \in \text{int}X_i$, $z \in \bigcup_j \text{int}X_j$.

Proof. $(a) \Rightarrow (b)$ Consider the multifunction F_μ as defined by (2.1). According to Theorem 2.1, μ is optimal if and only if the set $Q_0(\varphi_{F_\mu})$ is nonempty. Therefore, the implication will be proved if we show that nonemptiness of $Q_0(\varphi_{F_\mu})$ implies (b).

Since the family $\mathcal{X}$ is finite, it follows from (5.10) that

$$\text{spt}(\mu) = \bigcup_i \text{gr}(f^i),$$

where $\text{gr}(f^i) = \{(x,y) : x \in X_i, y = f^i(x)\}$ is the graph of f^i. Then

$$F_\mu(x) = \bigcup_{i \in I(x)} \{-c(\cdot, f^i(x))\},$$

where $I(x) := \{i : x \in X_i\}$. It follows from the definition of φ_{F_μ} (see(2.2)), that

$$\varphi_{F_\mu}(x,z) = \min_{i \in I(x)} \varphi_i(x,z) \quad \text{whenever } x,z \in X; \tag{5.13}$$

consequently,

$$\varphi_{F_\mu}(x,z) \le \varphi_i(x,z) \quad \text{whenever } x \in X_i, z \in X. \tag{5.14}$$

Let us take any $u \in Q_0(\varphi_{F_\mu})$ and define u_i to be the restriction of u to X_i. It follows from (5.14) that $u_i \in Q_0(\varphi_i|_{X_i \times X_i})$ and

$$\begin{aligned}\gamma(i,j) &= \min\{\varphi_i(x,z) - u(x) + u(z) : x \in X_i, z \in X_j\}\\ &\ge \min\{\varphi_{F_\mu}(x,z) - u(x) + u(z) : x \in X_i, z \in X_j\} \ge 0;\end{aligned}$$

therefore, (b) holds true with $\nu_i = 0$ for all i.

$(b) \Rightarrow (c)$ If (b) is satisfied, then, for any $x \in X_i \cap X_j$, we get from (5.12)

$$\begin{aligned}\nu_i - \nu_j &\le \varphi_i(x,x) - u_i(x) + u_j(x) = -u_i(x) + u_j(x),\\ \nu_j - \nu_i &\le \varphi_j(x,x) - u_j(x) + u_i(x) = -u_j(x) + u_i(x),\end{aligned}$$

hence $u_i(x) + \nu_i = u_j(x) + \nu_j$, and a function on X, $u(x) = u_i(x) + \nu_i$ for $x \in X_i, i \in I(x)$ is thus well-defined. Moreover, from (5.12) it follows that

$$u(x) - u(z) \le \varphi_i(x,z) \text{ for all } x \in X_i, z \in X,$$

therefore, the restriction of u to every X_i is continuous. Since all X_i are closed domains and $X = \bigcup_i X_i$, it follows that u is continuous on the whole X.

$(c) \Rightarrow (a)$ Since the set $\bigcup_i \text{int}X_i$ is dense in X and the function $\varphi_{F_\mu}(x,z) = \min\{\varphi_i(x,z) : i \in I(x)\}$ is continuous on $X \times X$, we derive from (c) that $u \in Q_0(\varphi_{F_\mu})$, and optimality of μ follows then from Theorem 2.1. □

Remark 5.1. If $Q_0(\varphi_i|_{X_i\times X_i})$ is nonempty, then, by Corollary 2.2, $\mu_{f^i}(\sigma_{1i})$ is an optimal solution to $MKP(c;\sigma_{1i},\sigma_{2i})$ and f^i is an optimal solution to $MP(c;\sigma_{1i},\sigma_{2i})$.

Remark 5.2. If $u_i \in Q_0(\varphi_i|_{X_i\times X_i})$, then $\gamma(i,i) = 0$, i.e., for $i = j$ (5.12) is satisfied automatically.

Theorem 5.4. *Suppose that* $\mathrm{spt}(\sigma_1) = X$. *If the family* $X = \{X_i\}$ *is countable, the set* $\bigcup_i \mathrm{int}X_i$ *is dense in* X, *its complement in* X *is* σ_1*-negligible, and there is a continuous function* u *on* X *such that* $u(x) - u(z) \le \varphi_i(x,z)$ *whenever* $x \in \mathrm{int}X_i$, $z \in \bigcup_j \mathrm{int}X_j$, *then* $\mu = \mu_f$ *is an optimal solution to* $MKP(c;\sigma_1,\sigma_2)$ *and* f *is an optimal solution to* $MP(c;\sigma_1,\sigma_2)$.

Proof. In virtue of Theorem 2.1, it suffices to show that $u \in Q_0(\varphi_{F_\mu})$. It follows from (5.10) that

$$\mathrm{spt}(\mu) = \overline{\bigcup_i \mathrm{gr}(f^i)},$$

where the line over $\bigcup_i \mathrm{gr}(f^i)$ means the closure of that set. Then, for any $x \in \mathrm{int}X_i$, $z \in X$,

$$\varphi_{F_\mu}(x,z) = \varphi_i(x,z),$$

hence

$$u(x) - u(z) \le \varphi_{F_\mu}(x,z). \tag{5.15}$$

In general case, we have

$$\varphi_{F_\mu}(x,z) = \inf_{y:(x,y)\in\mathrm{spt}(\mu)} (c(z,y) - c(x,y)) \tag{5.16}$$

(see (2.2)). It follows from (5.16) that φ_{F_μ} is upper semi-continuous on $X\times X$; therefore, the set $M := \{(x,z) : u(x) - u(z) \le \varphi_{F_\mu}(x,z)\}$ is closed in $X\times X$. Now, as $\bigcup_i \mathrm{int}X_i$ is dense in X and, in view of (5.15), $(\bigcup_i \mathrm{int}X_i) \times X \subset M$, we get $M = X \times X$, hence $u \in Q_0(\varphi_{F_\mu})$. □

6. Explicit solutions to Monge problems: examples and open questions

In this section we give several illustrative examples of explicite solutions to Monge problems. In all the examples we deal with the following simplest variant of the classic Monge problem. Given a cost function c and two convex bounded closed domains X and Y in $\mathbb{R}^n$ having the same volume (i.e., $\sigma_1 X = \sigma_2 Y$ where σ_1 and σ_2 stand for the n-dimensional Lebesgue measures on X and Y, respectively), the problem is to find a measure–preserving Borel map $f : X \to Y$, which is an optimal solution to the Monge problem

$MP(c;\sigma_1,\sigma_2)$. It follows from Theorems 1.2, 1.3, and 1.4 that for wide classes of cost functions c such a map exists and is unique up to values on a Lebesgue negligible set. For some other cost functions, an optimal measure–preserving map yet exists but is not unique. Basing on Theorem 2.1 and results of Section 5, we consider some cases when an optimal f can be produced explicitely. For the sake of simplicity, we restrict ourselves to the case $n = 2$.

In the first three examples, X is the square $\{x = (x_1, x_2) : |x_1| \le 1, |x_2| \le 1\}$ and Y is its shift, $Y = X + b$, where $b = (b_1, b_2)$ is an arbitrary vector in $\mathbb{R}^2$.

Example 6.1. If $c(x,y) = \|x-y\|^2 = (x_1-y_1)^2 + (x_2-y_2)^2$, then the shift $f(x) = x + b$ is the unique (up to values on a Lebesgue negligible set) optimal solution to the Monge problem $MP(c;\sigma_1,\sigma_2)$. This follows from Corollary 5.4. (Also this follows from Corollary 5.2 if one takes into account that $f(x) = -\nabla u(x)$ and $-u \in Q_0(\varphi)$, where $u(x) = -\frac{1}{2}x\cdot x - b\cdot x$.) Moreover, according to Corollary 5.4, this holds true for $c(x,y) = h(x-y)$ where h is any smooth (C^2) strictly convex function on $X - Y = \{z = x^1 - x^2 - b : x^1, x^2 \in X\}$.

Example 6.2. If $c(x,y) = \|x-y\| = \sqrt{(x_1-y_1)^2 + (x_2-y_2)^2}$ and $X \cap Y = \phi$, then $0 \notin X - Y = \{z = x^1 - x^2 - b : x^1, x^2 \in X\}$ and $h(z) = \|z\| = \sqrt{z_1^2 + z_2^2}$, $z = (z_1, z_2)$, is a smooth (C^2) convex function on $X - Y$. In this case, from Corollary 5.4 it follows that the shift $f(x) = x + b$ is an optimal solution to the Monge problem $MP(c;\sigma_1,\sigma_2)$. However, if $X \cap Y \ne \phi$, then $0 \in X - Y$, h fails to be smooth on $X - Y$, and Corollary 5.4 becomes inapplicable. Nevertheless, in this case the optimality of the shift $f(x) = x + b$ yet holds true. Indeed, consider the function $u(x) = -\frac{b\cdot x}{\|b\|}$. We have for every $x, z \in X$

$$\begin{aligned}
\varphi(x,z) - u(x) + u(z) &= \|z - x - b\| - \|b\| + \frac{b\cdot(z-x)}{\|b\|} \\
&= \|z - x - b\| - \|b\| + \frac{b\cdot(z-x-b)}{\|b\|} + \frac{b\cdot b}{\|b\|} \\
&= \|z - x - b\| - \|b\| + \frac{b}{\|b\|}\cdot(z-x-b) + \|b\| \\
&= \|z - x - b\| - \frac{(-b)}{\|b\|}\cdot(z-x-b) \ge 0,
\end{aligned}$$

that is, $u \in Q_0(\varphi)$. By Corollary 2.2, μ_f is an optimal solution to $MKP(c;\sigma_1,\sigma_2)$ hence f is an optimal solution to $MP(c;\sigma_1,\sigma_2)$. Let us show that the optimal solution need not be unique. To this end, we take $b = (1,0)$ and consider the map $f^1 : X \to Y$, $f^1(x) = x$ if $x \in X \cap Y = \{x = (x_1,x_2) : 0 \le x_1 \le 1, -1 \le x_2 \le 1\}$ and $f^1(x) = x + 2b$ if $x \in X \setminus Y = \{x = (x_1,x_2) : -1 \le x_1 < 0, -1 \le x_2 \le 1\}$. Clearly

$f^1 \in \Phi(\sigma_1, \sigma_2)$. Since the shift f is an optimal solution to $MP(c; \sigma_1, \sigma_2)$, we have

$$\mathcal{V}(c; \sigma_1, \sigma_2) = \int_X c(x, f(x))\, \sigma_1(dx) = \|b\| \sigma_1(X) = 4\|b\| = 4,$$

and as

$$\int_X c(x, f^1(x))\, \sigma_1(dx) = 2\|b\| \sigma_1(X \setminus Y) = 4\|b\| = 4,$$

we see that f^1 is optimal, as well.

Example 6.3. If $c(x, y) = -\|x - y\|^2 = -(x_1 - y_1)^2 - (x_2 - y_2)^2$, then the map $f(x) = -x + b$ is the unique (up to values on a Lebesgue negligible set) optimal solution to the Monge problem $MP(c; \sigma_1, \sigma_2)$. This follows from Corollary 5.5. Moreover, according to Corollary 5.5, this holds true for $c(x, y) = h(x - y)$ where h is any smooth (C^2) strictly concave function on $X - Y = \{z = x^1 - x^2 - b : x^1, x^2 \in X\}$.

In the following three examples, $b = (b_1, b_2) \in \mathbb{R}^2$,

$$\begin{aligned} X &= \{x = (x_1, x_2) : -2 \le x_1 \le 2, -1 \le x_2 \le 1\}, \\ X' &= \{x = (x_1, x_2) : (x_2, x_1) \in X\} \\ &= \{x = (x_1, x_2) : -1 \le x_1 \le 1, -2 \le x_2 \le 2\}, \\ Y &= X' + b \\ &= \{x = (x_1, x_2) : -1 + b_1 \le x_1 \le 1 + b_1, -2 + b_2 \le x_2 \le 2 + b_2\}. \end{aligned}$$

Example 6.4. If $c(x, y) = \|x - y\|^2 = (x_1 - y_1)^2 + (x_2 - y_2)^2$, then the affine map $f(x) = (f_1(x_1, x_2), f_2(x_1, x_2)) = (x_1/2 + b_1, 2x_2 + b_2)$ is the unique (up to values on a Lebesgue negligible set) optimal solution to the Monge problem $MP(c; \sigma_1, \sigma_2)$. This follows from Corollary 5.3 (I).

Example 6.5. If $c(x, y) = -\|x - y\|^2 = -(x_1 - y_1)^2 - (x_2 - y_2)^2$, then the affine map $f(x) = (f_1(x_1, x_2), f_2(x_1, x_2)) = (-x_1/2 + b_1, -2x_2 + b_2)$ is the unique (up to values on a Lebesgue negligible set) optimal solution to the Monge problem $MP(c; \sigma_1, \sigma_2)$. This follows from Corollary 5.3 (II).

Example 6.6. Suppose that $c(x, y) = \|x - y\|$ and $b = 0$ (hence $Y = X'$), and consider in $\mathbb{R}^2$ five closed convex domains X_i, $i = 1, \ldots, 5$, where

$$\begin{aligned} X_1 &= \{x = (x_1, x_2) : -1 \le x_1 \le 1, -1 \le x_2 \le 1\}, \\ X_2 &= \{x = (x_1, x_2) : 1 \le x_1 \le 2, 0 \le x_2 \le 1\}, \\ X_3 &= \{x = (x_1, x_2) : 1 \le x_1 \le 2, -1 \le x_2 \le 0\}, \\ X_4 &= \{x = (x_1, x_2) : -2 \le x_1 \le -1, -1 \le x_2 \le 0\}, \\ X_5 &= \{x = (x_1, x_2) : -2 \le x_1 \le -1, 0 \le x_2 \le 1\}. \end{aligned}$$

Clearly, $X = X_1 \cup X_2 \cup X_3 \cup X_4 \cup X_5$. Let us denote $e_1 = (1,0), e_2 = (0,1)$ and consider the shifts $f^i : X_i \to Y = X'$, where $f^1(x) = x$, $f^2(x) = x - e_1 + e_2$, $f^3(x) = x - e_1 - e_2$, $f^4(x) = x + e_1 - e_2$, and $f^5(x) = x + e_1 + e_2$. We take a map $f : X \to Y = X'$ that coincides with f^i on $\mathrm{int}X_i$, $i = 1, \ldots, 5$, and is defined arbitrarily on the complement of $\mathrm{int}X_1 \cup \mathrm{int}X_2 \cup \mathrm{int}X_3 \cup \mathrm{int}X_4 \cup \mathrm{int}X_5$ in X. We claim that f is an optimal solution to $MP(c; \sigma_1, \sigma_2)$. To show this, notice that $X \setminus (\mathrm{int}X_1 \cup \mathrm{int}X_2 \cup \mathrm{int}X_3 \cup \mathrm{int}X_4 \cup \mathrm{int}X_5)$ is a Lebesgue negligible Borel set; therefore,

$$\mathrm{spt}\,(\mu_f(\sigma_1)) = \bigcup_1^5 \mathrm{gr}(f^i)$$

and $\mu_f(\sigma_1) = \mu_{f^1}(\sigma_{11}) + \mu_{f^2}(\sigma_{12}) + \mu_{f^3}(\sigma_{13}) + \mu_{f^4}(\sigma_{14}) + \mu_{f^5}(\sigma_{15})$, where σ_{1i} stands for the Lebesgue measure on X_i, $i = 1, \ldots, 5$ (see (5.10) and Remark 5.2). It suffices to check that statement (b) of Theorem 5.3 holds true. To this end, we consider the functions φ_i on $X_i \times X$ as given by (5.11). A direct computation yields $\varphi_1(x,z) = \|z - x\|$, $\varphi_2(x,z) = \|z - x + e_1 - e_2\| - \sqrt{2}$, $\varphi_3(x,z) = \|z - x + e_1 + e_2\| - \sqrt{2}$, $\varphi_4(x,z) = \|z - x - e_1 + e_2\| - \sqrt{2}$, and $\varphi_5(x,z) = \|z - x - e_1 - e_2\| - \sqrt{2}$. Clearly, the set $Q_0(\varphi_1|_{X_1 \times X_1})$ is nonempty and the function

$$u_1(x) = -\frac{|x_1|}{\sqrt{2}} + \frac{|x_2|}{\sqrt{2}} + \nu_1 \tag{6.1}$$

is contained in it.The functions $\varphi_i|_{X_i \times X_i}, i = 2,3,4,5$, are smooth and convex.Then, by Corollary 5.4,$\mu_{f^i}(\sigma_{1i})$ is an optimal solution to $MKP(c; \sigma_{1i}, \sigma_{2i})$, $i = 2, \ldots, 5$, and, in virtue of Theorem 4.1, the sets $Q_0(\varphi_i|_{X_i \times X_i})$, $i = 2,3,4,5$, are nonempty. Moreover, by Theorem 3.1, each $u_i \in Q_0(\varphi_i|_{X_i \times X_i})$, $i = 2,3,4,5$ is unique up to a constant term ν_i and satisfies the equation $\nabla u_i(x) = \nabla_x \varphi_i(x,z)|_{z=x}$. Since

$$\nabla_x \varphi_i(x,z)|_{z=x} = -\frac{x - f^i(x)}{\|x - f^i(x)\|} = \frac{f^i(x) - x}{\sqrt{2}}, \quad i = 2,3,4,5,$$

we have

$$u_2(x) = -\frac{x_1}{\sqrt{2}} + \frac{x_2}{\sqrt{2}} + \nu_2, \tag{6.2}$$

$$u_3(x) = -\frac{x_1}{\sqrt{2}} - \frac{x_2}{\sqrt{2}} + \nu_3, \tag{6.3}$$

$$u_4(x) = \frac{x_1}{\sqrt{2}} - \frac{x_2}{\sqrt{2}} + \nu_4, \tag{6.4}$$

$$u_5(x) = \frac{x_1}{\sqrt{2}} + \frac{x_2}{\sqrt{2}} + \nu_5. \tag{6.5}$$

Now consider functions (6.1)–(6.5) with $\nu_i = 0$ and constitute the corresponding

$$\gamma(i,j) = \min\{\varphi_i(x,z) - u_i(x) + u_j(z) : x \in X_i, z \in X_j\}.$$

We have

$$\gamma(1,2) = \min\left\{\|z-x\| + \frac{|x_1|}{\sqrt{2}} - \frac{|x_2|}{\sqrt{2}} - \frac{z_1}{\sqrt{2}} + \frac{z_2}{\sqrt{2}} : x \in X_1, z \in X_2\right\},$$

$$\gamma(1,3) = \min\left\{\|z-x\| + \frac{|x_1|}{\sqrt{2}} - \frac{|x_2|}{\sqrt{2}} - \frac{z_1}{\sqrt{2}} - \frac{z_2}{\sqrt{2}} : x \in X_1, z \in X_3\right\},$$

$$\gamma(1,4) = \min\left\{\|z-x\| + \frac{|x_1|}{\sqrt{2}} - \frac{|x_2|}{\sqrt{2}} + \frac{z_1}{\sqrt{2}} - \frac{z_2}{\sqrt{2}} : x \in X_1, z \in X_4\right\},$$

$$\gamma(1,5) = \min\left\{\|z-x\| + \frac{|x_1|}{\sqrt{2}} - \frac{|x_2|}{\sqrt{2}} + \frac{z_1}{\sqrt{2}} + \frac{z_2}{\sqrt{2}} : x \in X_1, z \in X_5\right\},$$

$$\gamma(2,1) = \min\Bigg\{\|z-x+e_1-e_2\| \\ -\sqrt{2} + \frac{x_1}{\sqrt{2}} - \frac{x_2}{\sqrt{2}} + \frac{|z_1|}{\sqrt{2}} - \frac{|z_2|}{\sqrt{2}} : x \in X_2, z \in X_1\Bigg\},$$

$$\gamma(2,3) = \min\Bigg\{\|z-x+e_1-e_2\| \\ -\sqrt{2} + \frac{x_1}{\sqrt{2}} - \frac{x_2}{\sqrt{2}} - \frac{z_1}{\sqrt{2}} - \frac{z_2}{\sqrt{2}} : x \in X_2, z \in X_3\Bigg\},$$

$$\gamma(2,4) = \min\Bigg\{\|z-x+e_1-e_2\| \\ -\sqrt{2} + \frac{x_1}{\sqrt{2}} - \frac{x_2}{\sqrt{2}} + \frac{z_1}{\sqrt{2}} - \frac{z_2}{\sqrt{2}} : x \in X_2, z \in X_4\Bigg\},$$

$$\gamma(2,5) = \min\Bigg\{\|z-x+e_1-e_2\| \\ -\sqrt{2} + \frac{x_1}{\sqrt{2}} - \frac{x_2}{\sqrt{2}} + \frac{z_1}{\sqrt{2}} + \frac{z_2}{\sqrt{2}} : x \in X_2, z \in X_5\Bigg\}.,$$

$$\gamma(3,1) = \min\Bigg\{\|z-x+e_1+e_2\| \\ -\sqrt{2} + \frac{x_1}{\sqrt{2}} + \frac{x_2}{\sqrt{2}} + \frac{|z_1|}{\sqrt{2}} - \frac{|z_2|}{\sqrt{2}} : x \in X_3, z \in X_1\Bigg\},$$

$$\gamma(3,2) = \min\Bigg\{\|z-x+e_1+e_2\| \\ -\sqrt{2} + \frac{x_1}{\sqrt{2}} + \frac{x_2}{\sqrt{2}} - \frac{z_1}{\sqrt{2}} + \frac{z_2}{\sqrt{2}} : x \in X_3, z \in X_2\Bigg\},$$

$$\gamma(3,4) = \min\Big\{ \|z - x + e_1 + e_2\| - \sqrt{2} + \frac{x_1}{\sqrt{2}} + \frac{x_2}{\sqrt{2}} + \frac{z_1}{\sqrt{2}} - \frac{z_2}{\sqrt{2}} : x \in X_3, z \in X_4 \Big\},$$

$$\gamma(3,5) = \min\Big\{ \|z - x + e_1 + e_2\| - \sqrt{2} + \frac{x_1}{\sqrt{2}} + \frac{x_2}{\sqrt{2}} + \frac{z_1}{\sqrt{2}} + \frac{z_2}{\sqrt{2}} : x \in X_3, z \in X_5 \Big\},$$

$$\gamma(4,1) = \min\Big\{ \|z - x - e_1 + e_2\| - \sqrt{2} - \frac{x_1}{\sqrt{2}} + \frac{x_2}{\sqrt{2}} + \frac{|z_1|}{\sqrt{2}} - \frac{|z_2|}{\sqrt{2}} : x \in X_4, z \in X_1 \Big\},$$

$$\gamma(4,2) = \min\Big\{ \|z - x - e_1 + e_2\| - \sqrt{2} - \frac{x_1}{\sqrt{2}} + \frac{x_2}{\sqrt{2}} - \frac{z_1}{\sqrt{2}} + \frac{z_2}{\sqrt{2}} : x \in X_4, z \in X_2 \Big\},$$

$$\gamma(4,3) = \min\Big\{ \|z - x - e_1 + e_2\| - \sqrt{2} - \frac{x_1}{\sqrt{2}} + \frac{x_2}{\sqrt{2}} - \frac{z_1}{\sqrt{2}} - \frac{z_2}{\sqrt{2}} : x \in X_4, z \in X_3 \Big\},$$

$$\gamma(4,5) = \min\Big\{ \|z - x - e_1 + e_2\| - \sqrt{2} - \frac{x_1}{\sqrt{2}} + \frac{x_2}{\sqrt{2}} + \frac{z_1}{\sqrt{2}} + \frac{z_2}{\sqrt{2}} : x \in X_4, z \in X_5 \Big\},$$

$$\gamma(5,1) = \min\Big\{ \|z - x - e_1 - e_2\| - \sqrt{2} - \frac{x_1}{\sqrt{2}} - \frac{x_2}{\sqrt{2}} + \frac{|z_1|}{\sqrt{2}} - \frac{|z_2|}{\sqrt{2}} : x \in X_5, z \in X_1 \Big\},$$

$$\gamma(5,2) = \min\Big\{ \|z - x - e_1 - e_2\| - \sqrt{2} - \frac{x_1}{\sqrt{2}} - \frac{x_2}{\sqrt{2}} - \frac{z_1}{\sqrt{2}} + \frac{z_2}{\sqrt{2}} : x \in X_5, z \in X_2 \Big\},$$

$$\gamma(5,3) = \min\Big\{ \|z - x - e_1 - e_2\|$$
$$-\sqrt{2} - \frac{x_1}{\sqrt{2}} - \frac{x_2}{\sqrt{2}} - \frac{z_1}{\sqrt{2}} - \frac{z_2}{\sqrt{2}} : x \in X_5, z \in X_3 \Big\},$$
$$\gamma(5,4) = \min\Big\{ \|z - x - e_1 - e_2\|$$
$$-\sqrt{2} - \frac{x_1}{\sqrt{2}} - \frac{x_2}{\sqrt{2}} + \frac{z_1}{\sqrt{2}} - \frac{z_2}{\sqrt{2}} : x \in X_5, z \in X_4 \Big\}.$$

Tedious but elementary calculations show that $\gamma(i,j) \geq 0$ whenever $i \neq j$, and as $u_i \in Q_0(\varphi_i|_{X_i \times X_i})$, we see that $\gamma(i,i) \geq 0$ for every $i = 1, \ldots, 5$, as well. Then, by Theorem 5.3, μ_f is an optimal solution to $MKP(c; \sigma_1, \sigma_2)$ and f is an optimal solution to the Monge problem $MP(c; \sigma_1, \sigma_2)$.

In the following example, X and Y are congruent equilateral triangles: X is the triangle with vertexes $(-1,0), (1,0), (0,\sqrt{3})$ and $Y = -X$ is the triangle with vertexes $(1,0), (-1,0), (0,-\sqrt{3})$.

Example 6.7. Suppose that $c(x,y) = \|x - y\|$. We will consider quadrangles X_i determined as follows. Every positive integer i is represented uniquely in the form $i = 2^{k-1} + s$, where $s = s(i) \in \{0, \ldots, 2^{k-1} - 1\}$ and $k = k(i)$ is a positive integer. We define X_i to be the rhombus with vertexes $x^{i1} = (x_1^{i1}, x_2^{i1}), x^{i2} = (x_1^{i2}, x_2^{i2}), x^{i3} = (x_1^{i3}, x_2^{i3}), x^{i4} = (x_1^{i4}, x_2^{i4})$, where

$$x_1^{i1} = x_1^{i3} = -1 + \frac{1}{2^{k-1}} + \frac{s}{2^{k-2}},$$
$$x_1^{i2} = -1 + \frac{1}{2^k} + \frac{s}{2^{k-2}},$$
$$x_1^{i4} = -1 + \frac{3}{2^k} + \frac{s}{2^{k-2}},$$

$$x_2^{i1} = 0,$$
$$x_2^{i2} = x_2^{i4} = \frac{\sqrt{3}}{2^k},$$
$$x_2^{i3} = \frac{\sqrt{3}}{2^{k-1}}.$$

Clearly, $\mathrm{int}X_i \cap \mathrm{int}X_j = \phi$ whenever $i \neq j$, the set $\bigcup_{i=1}^{\infty} \mathrm{int}X_i$ is dense in X, and its complement in X is Lebesgue negligible. We consider the shifts $f^i : X_i \to Y$,

$$f^i(x_1, x_2) = (x_1, x_2 - \alpha_k),$$

where

$$\alpha_k := \frac{\sqrt{3}}{2^{k-1}}, \tag{6.6}$$

and take a map $f : X \to Y$ that coincides with f^i on $\mathrm{int}X_i$ for every i and is defined arbitrarily on the complement of $\bigcup_i \mathrm{int}X_i$ in X. We have

$$\mu_f(\sigma_1) = \sum_1^\infty \mu_{f^i}(\sigma_{1i}),$$

where σ_{1i} stands for the Lebesgue measure on X_i (see (5.10)). We claim that f is an optimal solution to $MP(c; \sigma_1, \sigma_2)$. This will follow from Theorem 5.4 if we show that there is a continuous function u on X such that $u(x) - u(z) \leq \varphi_i(x, z)$ whenever $x \in \mathrm{int}X_i$, $z \in \bigcup_j \mathrm{int}X_j$. A direct computation yields

$$\varphi_i(x, z) = \|z - f^i(x)\| - \|x - f^i(x)\| = \|z - x + \alpha_k e_2\| - \alpha_k$$
$$\forall x \in \mathrm{int}X_i, z \in \bigcup_j \mathrm{int}X_j,$$

where α_k is given by (6.6) and $e_2 = (0, 1)$. Let us consider on X the function $u(x) := -x_2$. We have for every $x \in \mathrm{int}X_i$, $z \in \bigcup_j \mathrm{int}X_j$,

$$\begin{aligned}\varphi_i(x, z) - u(x) + u(z) &= \|z - x + \alpha_k e_2\| - \alpha_k + x_2 - z_2 \\ &= \|z - x + \alpha_k e_2\| - (z - x + \alpha_k e_2) \cdot e_2 \geq 0,\end{aligned}$$

that is, u satisfies the assumption of Theorem 5.4. Now, applying the theorem establishes optimality of f.

In all previous examples, X and Y were congruent convex bodies in $\mathbb{R}^2$. In the next example, X and Y have the same area but are not congruent: X is the square with vertexes $(0,0), (0,1), (1,1), (1,0)$ and Y is the quadrangle with vertexes $(0,0), (1,1), (2,1), (1,0)$. We shall see that Theorem 5.3 is applicable to such situations, as well.

Example 6.8. Suppose that $c(x, y) = \|x - y\| = \sqrt{(x_1 - y_1)^2 + (x_2 - y_2)^2}$ and consider in $\mathbb{R}^2$ two closed convex domains, X_1 and X_2, where X_1 is the triangle with vertexes $(0,0), (1,1), (1,0)$ and Y is the triangle with vertexes $(0,0), (0,1), (1,1)$. Clearly, $X = X_1 \cup X_2$ and $\mathrm{int}X_1 \cap \mathrm{int}X_2 = \phi$. We define the maps $f^i : X_i \to Y$, $i = 1, 2$,

$$f^1(x) := x, \quad f^2(x) := x + e_1,$$

where $e_1 = (1, 0)$, and consider the map $f : X \to Y$ that coincides with f^i on $\mathrm{int}X_i$ for every i and is arbitrary on the complement of $\mathrm{int}X_1 \cup \mathrm{int}X_2$ in X. We claim that f is an optimal solution to $MP(c; \sigma_1, \sigma_2)$. To show this, notice that the complement of int$X_1 \cup$ intX_2 in X is a Lebesgue negligible Borel set; therefore, spt$(\mu_f(\sigma_1)) = $ gr$(f^1) \cup$ gr(f^2) and $\mu_f(\sigma_1) = \mu_{f^1}(\sigma_{11}) + \mu_{f^2}(\sigma_{12})$, where σ_{1i} stands for the Lebesgue measure on X_i (see (5.10)). A direct computation yields

$$\varphi_1(x,z) = \|z - f^1(x)\| - \|x - f^1(x)\| = \|z - x\|, \tag{6.7}$$

$$\varphi_2(x,z) = \|z - f^2(x)\| - \|x - f^2(x)\| = \|z - x - e_1\| - 1. \tag{6.8}$$

Clearly, the function $u_1(x) = x_1$ belongs to $Q_0(\varphi_1)$. The function $\varphi_2|_{X_2 \times X_2}$ is smooth and convex. Then, by Corollary 5.4, $\mu_{f^2}(\sigma_{12})$ is an optimal solution to $MKP(c; \sigma_{12}, \sigma_{22})$, where $\sigma_{22} = f^2(\sigma_{12})$ denotes the Lebesgue measure on the triangle Y_2 with vertexes $(1,0), (1,1), (2,1)$. In virtue of Theorem 4.1, the set $Q_0(\varphi_2|_{X_2 \times X_2})$ is nonempty. Moreover, by Theorem 3.1, each $u_2 \in Q_0(\varphi_2|_{X_2 \times X_2})$ is unique up to a constant term ν_2 and $\nabla u_2(x) = \nabla_x \varphi_2(x,z)|_{z=x}$. Since

$$\nabla_x \varphi_2(x,z)|_{z=x} = -\left.\frac{(z_1 - x_1 - 1, z_2 - x_2)}{\|e_1\|}\right|_{z=x} = (1,0) = e_1,$$

we conclude that $u_2(x) = x_1 + \nu_2$. Let us set $\nu_1 = \nu_2 = 0$. Taking into account (6.7) and (6.8), we get for every $j \in \{1,2\}$

$$\begin{aligned} \varphi_1(x,z) - u_1(x) + u_j(z) &= \|z - x\| - x_1 + z_1 \geq 0, \\ \varphi_2(x,z) - u_2(x) + u_j(z) &= \|z - x - e_1\| - 1 - x_1 + z_1 \\ &= \|z - x - e_1\| - (z - x - e_1) \cdot (-e_1) \geq 0; \end{aligned}$$

therefore, $\gamma(i,j) = \min\{\varphi_i(x,z) - u_i(x) + u_j(z) : x \in X_i, z \in X_j\} \geq 0$ whenever $i, j \in \{1,2\}$. Then, by Theorem 5.3, μ_f is an optimal solution to $MKP(c; \sigma_1, \sigma_2)$ and f is an optimal solution to the Monge problem $MP(c; \sigma_1, \sigma_2)$.

Also Example 6.8 can be examined basing on Theorem 1.1 for the space $X \cup Y$ with the Lebesgue measure on it if one takes into account that $\sigma_1 \wedge \sigma_2$ is the Lebesgue measure on $X \cap Y$.

Notice that Examples 6.1–6.5 and 6.8 are extended without change to broad classes of domains X and Y. In particular, if we consider Example 6.3 with X and Y from Example 6.7, then the map $f(x) = -x$ remains to be the unique (up to values on a Lebesgue negligible set) optimal solution to the Monge problem. Also, these examples are extended to convex bodies in $\mathbb{R}^n$ where $n > 2$. The main difficulty in concrete models is to guess what is an optimal map f. If f is guessed right, then results of Sections 4 and 5 can be used to establish its optimality.

In conclusion, we list several open questions that arise in connection with the above examples. In what follows, X and Y are convex figures of equal area in $\mathbb{R}^2$ with two-dimensional Lebesgue measures on them as σ_1 and σ_2. (Similar questions may be posed for convex bodies of equal volume in $\mathbb{R}^n$ where $n > 2$.) We will consider two types of the cost function: $c(x,y) = \|x-y\| = \sqrt{(x_1-y_1)^2+(x_2-y_2)^2}$ and $c(x,y) = \|x-y\|^2 = (x_1-y_1)^2+(x_2-y_2)^2$. If $c(x,y) = \|x-y\|^2$, then, by Theorem 1.2, there is a unique (up to values on a Lebesgue negligible set) optimal solution f to the Monge problem. If $c(x,y) = \|x-y\|$, then Theorem 1.2 becomes inapplicable. Two general questions arise as follows:

(Q_1) *Suppose* $c(x,y) = \|x-y\|^2$. *What is explicite form of optimal* f*?*

(Q_2) *Suppose* $c(x,y) = \|x-y\|$. *Is there an optimal solution to the Monge problem? If the answer is "yes", what is its explicite form for concrete* X *and* Y*?*

Let us specify X and Y and call several model variants of the questions as follows.

(Q_1):

1. X is as in Examples 6.1–6.3 and Y is the square with vertexes

$$(-\sqrt{2},0), (0,\sqrt{2}), (\sqrt{2},0), (0,-\sqrt{2}).$$

2. X and Y are as in Example 6.7.

3. X is a square and Y is a circle (the cases $X \cap Y = \phi$ and $X \cap Y \neq \phi$ are to be examined separately).

(Q_2):

1. X is as in Examples 6.1–6.3 and Y is the square X' with vertexes

$$(-\sqrt{2},0), (0,\sqrt{2}), (\sqrt{2},0), (0,-\sqrt{2})$$

or $Y = X' + b$, where $b = (b_1, b_2) \in \mathbb{R}^2$.

2. X is as in Example 6.7 and Y is the triangle $X' = -X + \frac{2}{3}\sqrt{3}e_2$ with vertexes

$$\left(-1, \frac{2}{3}\sqrt{3}\right), \left(1, \frac{2}{3}\sqrt{3}\right), \left(0, -\frac{1}{3}\sqrt{3}\right)$$

or $Y = X' + b$, where $\|b\|$ is large enough so that $X \cap Y = \phi$.

3. X is a square and Y is a circle (the cases $X \cap Y = \phi$ and $X \cap Y \neq \phi$ are to be examined separately).

4. X and Y are as in Examples 6.4–6.5 with $b \neq 0$.

References

[1] Brenier, Y.: Décomposition polaire et réarrangement monotone des champs de vecteurs. C.R. Acad. Sci. Paris Sér. I Math. **305**, 805–808 (1987)

[2] Brenier, Y.: Polar factorization and monotone rearrangement of vector-valued functions. Comm. Pure. Appl. Math. **44**, 375–417 (1991)

[3] Carlier, G., Ekeland, I., Levin, V.L., Shananin, A.A.: A system of inequalities arising in mathematical economics and connected with the Monge–Kantorovich problem. Ceremade–UMR 7534–Université Paris Dauphine, No 0213, 3 Mai 2002

[4] Evans, L.C.: Partial differential equations and Monge–Kantorovich mass transfer. In: Bott, R. et al., eds.: pp.26–78, Current Developments in Mathematics. International Press, Cambridge 1997

[5] Gangbo, W., McCann, R.J.: Optimal maps in Monge's mass transport problem. C.R. Acad. Sci. Paris Sér. I Math. **321**, 1653–1658 (1995)

[6] Gangbo, W., McCann, R.J.: The geometry of optimal transportation. Acta Math. **177**, 113–161 (1996)

[7] Kantorovich, L.V.: On mass transfer. Dokl. Akad. Nauk SSSR **37**(7–8), 199–201 (1942) (Russian)

[8] Kantorovich, L.V., Akilov, G.P.: Functional Analysis, 3rd ed. Moscow: Nauka 1984 (Russian)

[9] Kantorovich, L.V., Rubinshtein, G.S.: On a function space and some extremal problems. Dokl. Akad. Nauk SSSR **115**, 1058–1061 (1957) (Russian)

[10] Kantorovich, L.V., Rubinshtein, G.S.: On a space of countably additive functions. Vestnik Leningrad Univ. **13** (7), 52–59 (1958) (Russian)

[11] Levin, V.L.: Duality and approximation in the mass transfer problem. In: Mityagin, B.S. (ed): Mathematical Economics and Functional Analysis. Moscow: Nauka 1974, pp. 94–108 (Russian)

[12] Levin, V.L.: On the problem of mass transfer. Soviet Math. Dokl. **16**, 1349–1353 (1975)

[13] Levin, V.L.: Duality theorems in the Monge–Kantorovich problem. Uspekhi Matem Nauk **32** (3), 171–172 (1977) (Russian)

[14] Levin, V.L.: General Monge—Kantorovich problem and its applications in measure theory and mathematical economics. In: Leifman, L.J. (ed): Functional Analysis, Optimization, and Mathematical Economics. pp. 141–176, A Collection of Papers Dedicated to the Memory of L.V.Kantorovich. N.Y. Oxford: Oxford Univ. Press 1990a

[15] Levin, V.L.: A formula for optimal value of Monge-Kantorovich problem with a smooth cost function and characterization of cyclically monotone mappings. Matem. Sbornik **181**, N12, 1694–1709 (1990b)(Russian); English translation in Math. USSR Sbornik **71**, No 2, 533–548 (1992)

[16] Levin, V.L.: Some applications of set-valued mappings in mathematical economics. Journal of Mathematical Economics **20**, 69–87 (1991)

[17] Levin, V.L.: A superlinear multifunction arising in connection with mass transfer problems. Set-Valued Analysis **4**, N1, 41–65 (1996)

[18] Levin, V.L.: Reduced cost functions and their applications. Journal of Mathematical Economics **28**, 155–186 (1997a)

[19] Levin, V.L.: On duality theory for non-topological variants of the mass transfer problem. Matem. Sbornik **188**, N4, 95–126 (1997b)(Russian); English translation in Sbornik:Mathematics **188**, N4, 571–602 (1997)

[20] Levin, V.L.: Topics in the duality theory for mass transfer problems. In: Distributions with given marginals and moment problems (Beneš, V. and Štěpan, J., eds.), pp.243–252, Kluwer Academic Publishers 1997c

[21] Levin, V.L.: Existence and uniqueness of a measure preserving optimal mapping in a general Monge-Kantorovich problem. Funct. Anal. and Its Appl. **32**, No. 3, 205–208 (1998)

[22] Levin, V.L.: Abstract cyclical monotonicity and Monge solutions for the general Monge–Kantorovich problem. Set-Valued Analysis **7**, 7–32 (1999)

[23] Levin, V.L.: On generic uniqueness of optimal solutions for the general Monge–Kantorovich problem. Set-Valued Analysis **9**, 383–390 (2001a)

[24] Levin, V.L.: The Monge–Kantorovich problems and stochastic preference relations. Advances in Math. Economics **3**, 97–124 (2001b)

[25] Levin, V.L.: Optimality conditions for smooth Monge solutions of the Monge–Kantorovich problem. Funct. Anal. and Its Appl. **36**, No 2, 114–119 (2002)

[26] Levin, V.L.: Solving the Monge and the Monge–Kantorovich problems: theory and examples. Doklady Mathematics **67**, No 1, 1-4 (2003)

[27] Levin, V.L., Milyutin, A.A.: The problem of mass transfer with a discontinuous cost function and a mass statement of the duality problem for convex extremal problems. Russian Math. Surveys **34** (3), 1–78 (1979)

[28] Monge, G.: Mémoire sur la théorie des déblais et de remblais. Histoire de l'Académie Royale des Sciences de Paris, avec les Mémoires de Mathématique et de Physique pour la même année, 666–704 (1781)

[29] Rachev, S.T., Rüschendorf, L.: Mass Transportation Problems. Volume 1: Theory, Volume 2: Applications. Springer 1998

[30] Rüschendorf, L., Rachev, S.T.: A characterization of random variables with minimum L^2-distance. J. Multivariate Anal. **32**, 48–54 (1990)

[31] Sudakov V.N.: Geometric problems in the theory of infinite-dimensional probability distributions. Trudy MIAN **141** (1976)(Russian); English translation in: Proc. Steklov Inst. Math. **141** (1979)

Adv. Math. Econ. 6, 123–147 (2004)

Advances in
MATHEMATICAL
ECONOMICS

Valuation of mortgage-backed securities based on unobservable prepayment costs

Hidetoshi Nakagawa[1] **and Tomoaki Shouda**[2]

[1] Center for Research in Advanced Financial Technology, Tokyo Institute of Technology, 2-12-1 Ookayama, Meguro-ku, Tokyo, 152-8552, Japan
(e-mail: nakagawa@craft.titech.ac.jp)

[2] MTB Investment Technology Institute Co., Ltd., 2-5-6 Shiba, Minato-ku, Tokyo, 105-0014, Japan
(e-mail: shoda@mtec-institute.co.jp)

Received: July 8, 2003
Revised: October 28, 2003

JEL classification: C11, C13, G12

Mathematics Subject Classification (2000): 60 H 99

Abstract. Mortgage-backed securities (MBS) have been rapidly increasing in issued total amount recently in Japan.

In this paper, we consider the model where each loan borrower (mortgager) is specified only by a single variable called prepayment cost which is not directly observable in the secondary market. It is assumed that every mortgager does prepayment reasonably at the first time when the market interest rate decreases to the level that she or he thinks more advantageous to prepay the loan in consideration of her or his peculiar prepayment cost.

We study the theoretical MBS price in the arbitrage-free framework and present a procedure to estimate parameters of a distribution function of prepayment cost based on the history about riskless interest rate and actual prepayment events in terms of the Bayesian scheme.

Key words: mortgage-backed securities, prepayment risk, Bayesian model.

1. Introduction

RMBS (Residential Mortgage-backed securities) is a sort of coupon-bearing bond whose investors can receive cash-flows from a pool of many housing loans, or mortgages. While the housing loan is a comparatively excellent claim, the financial institution is always faced with the peculiar risk that some mortgagers fall to default and others may execute the option to pay the remaining principal of loan before super-long loan maturity. Therefore, the securitization of the housing loan like MBS has merit for both the investors who seek preferable financial assets and the financial institution that wants to hedge the loan risk.

In the United States, MBS market is a big bond market by which the Treasury bill market is followed in the market scale. On the other hand, though the MBS market is still developing in Japan, but the Housing Loan Corporation that has the maximum balance of money loans in Japan have accelerated the issue of MBS for a recent several years. MBS market is sure to mature in Japan in near the future.

As for MBS, unscheduled cash-flow may occur and so the principal decreases because of mortgagers' prepayment. The risk caused by such an uncertainty is called the prepayment risk. A mortgager may prepay because of refinancing another low-interest-rate loan when the interest rate decreases, another may sell her or his own house. Anyway, various are the reasons why mortgagers do prepayment and it is difficult to predict the happening of prepayment. In addition, attribute information on each mortgager who composes the pool is not open for the investor in MBS in general, owing to the anonymity of the loan claim. Only consolidated information like a monthly prepayment amount can be obtained if possible. Such conditions make it more difficult for investors to estimate the prepayment.

Therefore, the modeling of the prepayment risk of MBS is one of the important problems in the recent financing research and there are a lot of early researches on both theoretical and empirical sides.

In respect of practical applications, a proportional hazard model, which Schwartz and Torous [21] and so on proposed, has become popular. This is a kind of survival analysis; in a word, the method to use the historical data to estimate the prepayment probability at the fixed times by modeling the yield curve at that time and the attribute of mortgage pool and so on as state variables. In this model, the term structure of prepayment probability is represented by a deterministic function called a base line hazard. Ichijou and Moridaira [8] analyze the model of this type for a housing loan in Japan. Deng and Quingley [1] study unobservable heterogeneity of mortgagers in the framework of proportional hazard model.

The formulation of prepayment risk as an application of the option theory has been studied by Dunn and McConell [4] [5]. They formulate the prepayment risk under the assumption that each mortgager acts reasonably to exercise her or his option to prepay at the optimal time. In Dunn and Spatt [6], a realistic assumption was put that each mortgager has different prepayment cost, or prepayment penalty. Stanton [22] assumed that the distribution of the prepayment cost among mortgagers is given by a beta distribution and estimated the parameters from actual GNMA prepayment data. Kariya and Kobayashi [11] introduce a notion of interest incentive function so as to describe the heterogeneity of the incentives of the borrowers for refinancing. In the discrete-time setting, the prepayment time is defined by the first time when the decline of the borrower's mortgage rate from the initial rate is over each mortgager's incentive threshold. Kariya et al. [12] considers that a fall in the housing price level from its initial level as well as a decline of interest rate may trigger prepayments. Basically, such ideas and prepayment cost are very similar in the sense that both would explain the burnout through the heterogeneity of mortgagers. As for rational prepayment model, see also Richard and Roll [20] and Mcconnell and Singh [16].

In Nakamura [18], the above models are called structural approach models by analogy of the default risk valuation model. On the other hand, some models are called reduced-form approach models; the models take the approach to formulate the occurrence of prepayment event as a stochastic event independent of interest rate by modeling the probability of instantaneous prepayment as a stochastic process called hazard rate. Nakamura [18] and Shibasaki and Nakamura [19] has proposed an efficient valuation method in comparison with the Monte Carlo method used in general respectively for both a reduced-form approach model and a structural approach model.

This paper proposes a new model of prepayment risk based on the structural approach under the presumption that each mortgager has different prepayment cost and acts reasonably to exercise her or his option to prepay at the optimal time. In our model, prepayment costs in the same pool are supposed to be independently chosen from a parametric probability distribution function. This implies that the investors cannot learn the true values of prepayment costs until an actual prepayment happens, and still they cannot grasp who prepaid due to the privacy protection. This point is essentially different from the default risk analysis. It is quite difficult to analysis the relation between prepayment and each mortgager's own characteristics while it is common to analysis the relation between default event and the financial statement of each firm.

To make the argument simple, it is assumed that all mortgagers have the same attributes except prepayment cost. In other words, it is considered that each mortgager's private information (age, income, region and so on) is transformed and collected into a form of prepayment cost. Moreover, any time-dependent exogenous state variables other than interest rates are not considered.

We also ignore the possibility that some mortgagers fall into default or partial prepayment. This means that only full prepayment causes mortgagers to leave the mortgage pool.

MBS is defined as the pass-through security whose cash-flow from mortgagers in the pool is directly paid to the investor. In addition, we remark that although we adopt a continuous-time model, as is different from most of the literature about MBS, it is not hard to reduce our model to a common discrete-time (monthly) framework so that one can estimate parameters in the model by using actual data.

We think about the problem of pricing of MBS in the arbitrage-free framework as well as estimating the parameter which decides the distribution of prepayment cost from only information observed in the riskless interest rate market and the number of announced prepayment. Hence, our result can explicitly explain the burnout effect of prepayment.

This paper is organized as follows. First of all, we formulate in section 2 the interest rate market, mortgage pool and the distribution of prepayment cost. In section 3, the theoretical price of MBS is derived under the condition that the parameter of the prepayment cost distribution is given. Moreover, we give the stochastic differential equation followed by the process of MBS price when the parameter of the prepayment cost distribution is given. This helps to comprehend that the variation of each factor affects the variation of the MBS price. In Section 4, we pay attention to the conditional density of the parameter of the prepayment cost distribution given the history of riskless interest rate and prepayment occurrences. The Bayesian approach is utilized there. Section 5 presents a simple example and the way of parameter estimation from the observation data. Section 6 gives the conclusion. In the appendix, the proofs of results in Section 3 is given.

The authors express gratitude to anonymous referees for courteous and useful comments.

2. The model

2.1 The model of interest rate market

First, we describe an interest rate market. Let $(\Omega^r, \mathcal{G}^r, P^r)$ be a complete probability space and T_0 be a finite time horizon. We denote by $(B_t^P)_{0 \le t \le T_0}$ a one-dimensional Brownian motion defined on $(\Omega^r, \mathcal{G}^r, P^r)$.

We suppose that there exists a risk-free short rate process $r(t)$ satisfying the following stochastic differential equation,

$$dr(t) = \mu^P(r(t), t)dt + \sigma^P(r(t), t)dB_t^P, \tag{1}$$

where $\mu^P(x, t)$ and $\sigma^P(x, t)$ are deterministic functions which satisfy some regularity conditions for the existence of unique strong solution of (1).

Denote by $(\mathcal{G}_t^r)_{0 \le t \le T_0}$ be the (right continuous) filtration defined by

$$\begin{aligned}\tilde{\mathcal{G}}_t^r &= \sigma\{r(s), s \le t\} \; (= \sigma\{B_s^P, s \le t\}) \\ \mathcal{G}_t^r &= \bigcap_{s>t} \tilde{\mathcal{G}}_s^r.\end{aligned}$$

This may be interpreted as the information observed in the interest rate market up to the time t.

Besides, we suppose that there exists a risk-neutral probability measure Q^r for the interest rate market so that $P(t,T)$, the price at time t of default-free zero coupon bond with maturity T, could be represented as

$$P(t,T) = E^Q\Big[\exp\Big(-\int_t^T r(u)du\Big)\Big|\mathcal{G}_t^r\Big],$$

where $E^Q[\cdot]$ stands for the expectation under the probability Q.

Since there exists an adapted process $\lambda(t)$ such that

$$E^P\Big[\frac{dQ}{dP}\Big|\mathcal{G}_t^r\Big] = \exp\Big(\int_0^t \lambda(u)dB_u^P - \frac{1}{2}\int_0^t \lambda(u)^2du\Big)\text{a.s.}$$

Then, under the risk-neutral measure Q^r, $r(t)$ satisfies

$$dr(t) = \mu^Q(r(t),t)dt + \sigma^P(r(t),t)dB_t^Q, \tag{2}$$

where

$$\begin{aligned}B_t^Q &= B_t^P - \int_0^t \lambda(s)ds \\ \mu^Q(x,t) &= \mu^P(x,t) + \sigma^P(x,t)\lambda(t)\end{aligned}$$

B_t^Q is a Q^r-Brownian motion due to Girsanov-Maruyama theorem. (For example, refer to Chapter 6 of Lamberton and Lapayre [14] for details.)

2.2 The model of mortgage pool

First, we present the presumptions when constructing a prepayment model later.

- In the mortgage pool, every loan follows the same conditions about the initial principal, the mortgage loan, the term of repayment and so on.
- Every mortgager is assumed to have the same characteristics except a prepayment cost which is finely described later. In other words, every (time-independent) personal information is transformed and collected into a form of prepayment cost, and any time-dependent exogeneous state variables other than interest rates are not considered. The sensitivity of each mortgager to the interest rate depends on her or his prepayment cost.

- Every mortgager is rational in the sense that she or he surely prepay the loan when it is judged that prepayment including the accompanying cost is profitable to holding the mortgage. However, they are not assumed to be risk-neutral because they tries to optimize their future debt, not under the risk-neutral measure Q, but under the original measure P.
- Partial prepayment is not allowed in the paper for simplicity. That is, if a mortgager likes to redeem her or his loan, she or he must prepay a full remaining principal at a time. As a matter of fact, especially in Japanese mortgage market where commissions for partial prepayment is relatively low and partial prepayment is not negligible. In the future, we will consider the prepayment model containing partial prepayment cases. Besides we realize that it is necessary to take care of handling the data which include partial prepayment case at the empirical analysis.

We consider that the mortgage pool consists of N mortgagers and the distribution of prepayment cost includes d parameters (see in detail later).

Let $\mathcal{B}^k \equiv \mathcal{B}(\mathbf{R}^k)$ for a positive integer k and let $(\Omega^b, \mathcal{G}^b) = (\mathbf{R}^N \times \mathbf{R}^d, \mathcal{B}^N \times \mathcal{B}^d)$ be a product measurable space.

We also define an enlarged probability space by considering a probability measure P^b on $(\Omega^b, \mathcal{G}^b)$,

$$(\Omega, \mathcal{F}, P) = (\Omega^r \times \Omega^b, \mathcal{G}^r \otimes \mathcal{G}^b, P^r \times P^b).$$

Let us suppose that the characterization of sinking-fund bond for each mortgager is given as follows.

- $c(t)$ stands for the scheduled coupon rate continuously paid by a mortgager at time t. (given)
- R is the fixed mortgage rate which is specified by the contract. (given)
- $T(< T_0)$ is the maturity of the loan. (given)
- $p(t)$ means the outstanding principal of the loan at time t given by

$$p(t) = \int_t^T e^{-R(s-t)} c(s) ds \tag{3}$$

We remark that $p(t)$ is a unique solution to the following ordinary differential equation

$$\frac{dp(t)}{dt} = Rp(t) - c(t), \qquad p(T) = 0. \tag{4}$$

On the other hand, we assume that each mortgager is specified respectively through an idea of prepayment cost. The followings are our fundamental assumptions on the prepayment cost.

- In general, a prepayment cost is recognized as the total additional cost except for the principal at prepayment, including some kinds of commission or service fees, a deposit for guarantee and so on. However, we will view it as the measure of ease for prepayment rather than the absolute cost of prepayment exercise though we try to consider the aspect of absolute cost later.

- Actual prepayment costs are supposed to be given at the beginning and constant in time. However, when we see this matter from an MBS investor's viewpoint, the prepayment costs in the pool cannot be directly measured and there are few available data to analyze the prepayment behavior minutely. Thus we assume that the prepayment costs are modeled as real valued random variables and that they are independent and identically distributed with some parameter. At last, we try to estimate this parameter from the available data such as interest rate and monthly report on prepayment.
- We infer that the level of interest rate relative to the mortgage rate has more influence on the prepayment behavior of the mortgagers than other covariates. In the sense, it seems natural to think that the prepayment cost depends on interest rates (and time). However, we suppose that prepayment costs are constant as is described above. This is why we suppose that the interest rate has an influence not on prepayment costs directly but the parameter of the distribution function. This means that the MBS investors learn to know how the prepayment costs are actually distributed through the estimated parameter from the interest rate.
- The most important thing to price the MBS in our study is to obtain the better information on how prepayment costs are distributed in the underlying pool, not to consider the exact relation between prepayment cost and interest rate. Therefore our presumptions are relevant at the implementation.

Here we prepare for some notation.

- $b_i \in \mathbf{R}(i = 1, \cdots, N)$ is an independent and identically distributed random variable that is interpreted as a prepayment cost for a mortgager i. Prepayment cost is considered more in the next subsection. Assume that $P(b_i = b_j) = 0$ for $i \neq j$. (That is, prepayment costs of all the mortgagers are different.)
- $\tau(b)$ is a $(\mathcal{G}_t^\tau)$-stopping time interpreted as the prepayment time for mortgager with prepayment cost b. Its characterization is done later.
- $N(t, b)$ represents the process counting whether a mortgager with prepayment cost b has already prepaid by time t defined by

$$N(t, b) = 1_{\{\tau(b) \leq t\}}.$$

- $D(t, T, b)$ is the total amount that is paid to MBS investors if a mortgager with prepayment cost b repays her or his loan from time t to either T or $\tau(b)$. $D(t, T, b)$ is given by

$$D(t, T, b) = \int_t^T Z(t, s)(1 - N(s, b))c(s)ds + \int_t^T Z(t, s)p(s)dN(s, b),$$

where

$$Z(t,s) = \exp\Big(-\int_t^s r(u)du\Big).$$

We also assume that the investor in MBS cannot directly observe $b_1, \cdots, b_N$ as well as which mortgager has prepaid. However, we suppose that it is possible to observe the total number of mortgagers who have prepaid up to time t, $A(t)$, given by

$$A(t) = \sum_{i=1}^{N} N(t, b_i).$$

Here, we define a new filtration $\mathcal{H}_t^A = \sigma\{A(s); s \leq t\}$. In addition, we define another filtration $(\mathcal{F}_t)$ by

$$\mathcal{F}_t = \bigcap_{s>t} \{\mathcal{G}_s^r \vee \mathcal{H}_s^A\}.$$

This filtration can be regarded as information about not only the interest rate market but when prepayment happened.

Remark 2.1. The cash-flow of MBS during the small period $(t, t+h]$ is approximated by

$$\begin{aligned} &\sum_{i=1}^{N} \{(1 - N(t,b_i))c(t)h + p(t+h)(N(t+h,b_i) - N(t,b_i)\} \\ &\quad = \quad (N - A(t))c(t)h + p(t+h)(A(t+h) - A(t)). \end{aligned}$$

Furthermore, this can be reduced by using (4) to

$$-N \cdot p(t+h)\frac{N - A(t+h)}{N} + N \cdot \{p(t) + Rp(t)h\}\frac{N - A(t)}{N}.$$

This representation of cash-flow approximation just corresponds to the discrete-time cash-flow function in section 2 in Kariya and Kobayashi [11]. This implies that our model is consistent with the common discrete-time model.

We remark that every actual prepayment cost less than a certain prepayment cost becomes $\mathcal{F}_t$-measurable as time passes in the following sense.

Define a $\mathbf{R}$-valued non-decreasing process $U(t)$, called the upper bound of prepayment cost up to time t, by

$$U(t) = \sup\{b \in \mathbf{R} | \tau(b) \leq t\}.$$

This means that every mortgager with prepayment cost which is below $U(t)$ has prepaid by time t, which this shows a so-called burnout effect. On the other hand, mortgagers with prepayment cost above $U(t)$ never prepay at time t.

2.3 Prepayment cost and optimal prepayment threshold

Rigorously, we should use a word of prepayment cost in the sense of absolute cost of prepayment action. The point is that a rational mortgager ought to prepay as soon as the sum of the remaining principal and her or his prepayment cost is smaller than or equal to the present value of the expected total amount of future repayment, in short, a lump-sum prepayment is preferable to the continuation of repayment.

Naively, this is reduced to the following optimal stopping problem due to the Markov property of $r(t)$ given gy (1):

$$\begin{aligned} &G(t,r,b) \\ &= \inf_{\tau\in\mathcal{T}_t} E^P\Big[\int_t^{\tau\wedge T} Z(t,s)c(s)ds + Z(t,\tau)\{p(\tau)+b\}1_{\{\tau<T\}}|r(t)=r,b\Big], \end{aligned}$$

where $\mathcal{T}_t$ denote the class of $(\mathcal{G}_t^{r,b})$-stopping times which take value more than t.

It can be checked that $G(t,r,b)$ is non-increasing in r by the argument based on the comparison theorem. Moreover, remark that $G(t,r,b)$ is non-decreasing in b, and for any $t<T$ and any b,

$$G(t,r,b) \le p(t)+b.$$

Let us define a deterministic function $S(t,b)$ as

$$S(t,b) = \sup\{r\in\mathbf{R}|G(t,r,b)=p(t)+b\}. \quad (\sup\emptyset = -\infty) \tag{5}$$

From a general discussion on the optimal stopping theorem, it follows that $S(t,b)$ gives an optimal prepayment threshold of riskless interest rate in the sense that a rational mortgager with prepayment cost b prepays at time t if the interest rate falls down to or below $S(t,b)$.

If $r(t)$ is nonnegative for any t, we can see that for $b'\le b$ and any $\varepsilon>0$,

$$\begin{aligned} &G(t,S(t,b)-\varepsilon,b') \\ &= \inf_{\tau\in\mathcal{T}_t} E^P\Big[\int_t^{\tau\wedge T} Z(t,s)c(s)ds + Z(t,\tau)\{p(\tau)+b'\}1_{\{\tau<T\}}| \\ &\qquad\qquad r(t)=S(t,b)-\varepsilon,b,b'\Big] \\ &= \inf_{\tau\in\mathcal{T}_t}\Big\{E^P\Big[\int_t^{\tau\wedge T} Z(t,s)c(s)ds + Z(t,\tau)\{p(\tau)+b\}1_{\{\tau<T\}}| \\ &\qquad\qquad r(t)=S(t,b)-\varepsilon,b,b'\Big] \\ &\qquad -E^P[Z(t,\tau)(b-b')1_{\{\tau<T\}}|r(t)=S(t,b)-\varepsilon,b,b']\Big\} \\ &\ge G(t,S(t,b)-\varepsilon,b)-(b-b'). \end{aligned}$$

Then the equality holds when $\tau \equiv t$ and it follows from (5) that

$$G(t, S(t,b) - \varepsilon, b') = p(t) + b - (b - b') = p(t) + b'.$$

By arbitrariness of ε, this leads to

$$S(t,b) \leq S(t,b') \text{ if } b \geq b'. \tag{6}$$

That is, the higher a mortgager's prepayment cost is, the lower the threshold of riskless interest rate which gives her or him the incentive to prepayment is.

The consideration above implies that $S(t,b)$ is specified endogenously by the coupon rate $c(t)$ and the fixed rate R (through $p(t)$) as well as time and prepayment cost. However, it is so hard to specify $S(t,b)$ exactly as is seen from pricing the American derivatives. Therefore we just think of prepayment cost as the covariate that gives the optimal level of interest rate for prepayment. Roughly, this states that mortgagers do not necessarily know the optimal level of interest rate for prepayment in the strict sense, but all of them just believe that $S(t,b)$ is given and it is the optimal threshold for time t and prepayment cost b of interest rate which gives the incentive to prepayment to a mortgager with prepayment cost b.

Hereafter we suppose that $S(t,b)$ is exogenously given so that it becomes so close to the function specified by (5) as possible and satisfies the following good properties, even though it might not be rational in the rigorous sense.

Assumption 2.2. *$S(t,b)$ is continuous in both t and b and satisfies the strict inequality on* (6)*, that is, we have $S(t,b) < S(t,b')$ for $b > b'$.*

If such $S(t,b)$ is given, we specify the prepayment time $\tau(b)$ by

$$\tau(b) = \inf\{t > 0 | r(t) \leq S(t,b)\}. \tag{7}$$

Then the upper bound of prepayment cost $U(t)$ can be rewritten as follows.

$$U(t) = \sup\{b \in \mathbf{R} | \min_{0 \leq u \leq t} \{r(u) - S(u,b)\} \leq 0\}, \tag{8}$$

Then we show the following claim.

Lemma 2.3. *Under the above assumptions, $U(t)$ is a $(\mathcal{G}_t^{\tau})$-adapted and continuous.*

Proof. In order to show the left continuity of $U(t)$, suppose there exist some $s \leq T$ and $\delta > 0$ such that $U(s) - U(s-) \geq \delta$ with positive probability, where $U(s-) = \lim_{u \uparrow s} U(u)$. ($U(t)$ is non-decreasing, so the left limit exists.)

Then for any $u < s$ and any $\varepsilon > 0$, we have $r(u) > S(u, U(s-) + \varepsilon)$, hence in particular, $r(u) > S(u, U(s))$.

As a consequence, the assumption that $r(u)$ is continuous and that $S(s, \cdot)$ is strictly decreasing implies that

$$r(s) = \lim_{u \uparrow s} r(u) \geq \lim_{u \uparrow s} S(u, U(s-)) = S(s, U(s-)) > S(s, U(s)).$$

However, it must hold that $S(s, U(s)) \geq r(s)$ due to the definition of $U(s)$. This leads to the contradiction. A similar argument can be applied to the right continuity.

2.4 The distribution of prepayment cost

As is mentioned above, we assume that individual prepayment costs of mortgagers $b_1, \ldots, b_N$ are impossible for MBS investors to observe directly and are looked upon as independent samples of N pieces of some random variables.

Assume that there exists a probability measure Q^b equivalent to P^b on the measurable space $(\Omega^b, \mathcal{F}^b)$ such that under a risk neutral probability measure Q on $(\Omega, \mathcal{F})$ defined as $Q = Q^r \times Q^b$, the theoretical price of pass-through type MBS is given by

$$X(t) = E^Q \Big[\sum_{i=1}^{N} D(t, T, b_i) \Big| \mathcal{F}_t \Big]. \tag{9}$$

In order to calculate this expectation, we need to specify $b_1, \cdots, b_N$. Recalling the fundamental assumptions described in subsection 2.2, we pay attention to the parametric distribution of prepayment costs, especially under the risk-neutral probability Q.

Under Q^b, we suppose that there is a distribution function $F(x;\theta)$ with an d-dimensional vector θ as parameter, that is,

$$F(x;\theta) = Q^b(b_i \leq x|\theta),$$

for any $i = 1, \cdots, N$.

Hence because of independence of prepayment costs we have

$$Q^b(b_1 \leq x_1, \cdots, b_N \leq x_N) = \int_{\mathbf{R}^d} \prod_{i=1}^{N} F(x_i;\theta)\psi(\theta) d\theta,$$

where $\psi(\theta)$ is a prior density of θ under Q^b.

The point is that θ is unknown parameter to the MBS market participants. (Perhaps the market originator may know the true value.) We thus give a prior density of θ and investigate the posterior density $Q(\theta|\mathcal{F}_t)$ of θ given the market data up to time t based on Bayes scheme, which is explained in section 4.

As a remark, we implicitly argue that it is not the prepayment costs in themselves but the post distribution of prepayment costs that changes in time. Someone may criticize that it sounds strange to assume that prepayment cost is unchangeable to each mortgager and independent of the information on future's interest rate dynamics. However it is quite difficult for the MBS investor to grasp the whole of the mortgage pool and we think it useful to simplify the situation and transfer the issue from prepayment costs to the distribution of them. Therefore we tried to formulate such a distribution with a parameter estimated by the path of interest rate movement and the observed actual prepayment. Indeed, our next concern is the model of prepayment cost dependent upon the movement of interest rate.

For the original measure P, we assume that the distribution function of prepayment cost is the same as that with respect to Q, where the parameter may be different.

3. Pricing of MBS when θ is known

Throughout this section, we assume that the parameter θ for the distribution of prepayment cost under Q is known. Let us denote the filtrations by $\mathcal{G}_t^{r,\theta}$ and $\mathcal{F}_t^\theta$ in order to make it apparent. Moreover, $F(x;\theta)$ is assumed to be continuously differentiable and denote the density function by $f(x;\theta)$.

Under the risk-neutral probability measure Q, the theoretical price of pass-through type MBS for the mortgage pool specified in the last section is given by the following expression.

$$
\begin{aligned}
X(t;\theta) &= E^Q\Big[\sum_{i=1}^N D(t,T,b_i)\Big|\mathcal{F}_t^\theta\Big] \\
&= E^Q\Big[\sum_{i=1}^N \Big\{\int_t^T Z(t,s)(1-N(s,b_i))c(s)ds \\
&\qquad + \int_t^T Z(t,s)p(s)dN(s,b_i)\Big\}\Big|\mathcal{F}_t^\theta\Big] \\
&= E^Q\Big[\int_t^T Z(t,s)(N-A(s))c(s)ds \\
&\qquad + \int_t^T Z(t,s)p(s)dA(s)\Big|\mathcal{F}_t^\theta\Big] \\
&= \int_t^T c(s)E^Q[Z(t,s)(N-A(s))|\mathcal{F}_t^\theta]ds \\
&\qquad + E^Q\Big[\int_t^T Z(t,s)p(s)dA(s)\Big|\mathcal{F}_t^\theta\Big]. \quad (10)
\end{aligned}
$$

Letting $\Phi(t) = Q(\tau(b) \leq t | \mathcal{G}_t^{r,\theta})$, the following result holds.

Lemma 3.1.

$$\Phi(t) = F(U(t); \theta)$$

Proof. It immediately follows from $\{b \in \mathbf{R} | \tau(b) \leq t\} = \{b \in \mathbf{R} | b \leq U(t)\}$.

Next, we define

$$\Gamma(t) = -\log(1 - \Phi(t)). \tag{11}$$

This equality is equivalent to

$$1 - \Phi(t) = e^{-\Gamma(t)}.$$

We assume $E^Q[\Gamma(T)] < \infty$ hereafter.

One can see that both $\Phi(t)$ and $\Gamma(t)$ are continuous processes due to the continuity of both $F(\cdot; \theta)$ and $U(t)$.

The following is our main result in this section.

Theorem 3.2. *The theoretical price of MBS when θ is given, denoted by $X(t; \theta)$, is given by*

$$X(t;\theta) = (N - A(t))e^{\Gamma(t)} E^Q\left[\int_t^T Z(t,s)e^{-\Gamma(s)}[c(s)ds + p(s)d\Gamma(s)]\middle| \mathcal{G}_t^{r,\theta}\right].$$

Specifically, in terms of θ,

$$\begin{aligned} X(t;\theta) \quad = \quad & \frac{N - A(t)}{1 - F(U(t);\theta)} \\ & \times \left\{ E^Q\left[\int_t^T Z(t,s)(1 - F(U(s);\theta))c(s)ds \middle| \mathcal{G}_t^{r,\theta}\right] \right. \\ & \left. + E^Q\left[\int_t^T Z(t,s)p(s)f(U(s);\theta)dU(s) \middle| \mathcal{G}_t^{r,\theta}\right] \right\}. \end{aligned} \tag{12}$$

This theorem is proved in Appendix.

Based on the result in Theorem 3.2, we derive the stochastic differential equation that the theoretical price $X(t)$ satisfies. The SDE expression enables us to understand how much the movement of each variable has an impact on the movement of MBS price easily.

Let

$$\tilde{V}(t) = E^Q\left[\int_t^T Z(t,s)e^{-\Gamma(s)}[c(s)ds + p(s)d\Gamma(s)]\middle| \mathcal{G}_t^{r,\theta}\right]$$

and $V(t) = Z(0,t)\tilde{V}(t)$. Then we have

$$\begin{aligned} V(t) &= E^Q\left[\int_0^T Z(0,s)e^{-\Gamma(s)}[c(s)ds + p(s)d\Gamma(s)]\middle|\mathcal{G}_t^{r,\theta}\right] \\ &\quad - \int_0^t Z(0,s)e^{-\Gamma(s)}[c(s)ds + p(s)d\Gamma(s)]. \end{aligned} \tag{13}$$

Since the fist term in right-hand is a square-integrable $(Q, (\mathcal{G}_t^{r,\theta}))$-martingale, the martingale representation theorem for Brownian filtration implies that there exists a $(\mathcal{F}_t^\theta)$-predictable process $\alpha(t)$ with

$$E^Q\left[\int_0^T \alpha(s)^2 ds\right] < \infty$$

such that

$$\begin{aligned} &E^Q\left[\int_0^T Z(0,s)e^{-\Gamma(s)}[c(s)ds + p(s)d\Gamma(s)]\middle|\mathcal{G}_t^{r,\theta}\right] \\ &\quad = V(0) + \int_0^t V(s)\alpha(s)dB_s^Q. \end{aligned} \tag{14}$$

In this case, we have $\alpha(t)dt = \frac{1}{V_t} d\langle V, B^Q\rangle_t$, so $\alpha(t)$ is approximately obtained by computation of the bracket.

Then we have

Proposition 3.3. *The theoretical price of MBS satisfies the following stochastic differential equation:*

$$\begin{aligned} &dX(t;\theta) \\ = &\Big(X(t;\theta)\big(r(t) - \alpha(t)\sigma(r(t),t)^{-1}\mu^Q(r(t),t)\big) - (N - A(t))c(t)\Big)dt \\ &+X(t;\theta)\alpha(t)\sigma(r(t),t)^{-1}dr(t) \\ &-(N - A(t-))^{-1}X(t-;\theta)dA(t) \\ &+\Big(X(t;\theta) - (N - A(t))p(t)\Big)\tilde{f}(U(t);\theta)dU(t), \end{aligned}$$

where

$$\tilde{f}(U(t);\theta) = \frac{f(U(t);\theta)}{1 - F(U(t);\theta))}.$$

As above, the variation of arbitrage-free price of MBS can be decomposed into the following causes — the time-varying part, the variation of riskless interest rate and the increase in the total number of prepayments and in the upper bound of prepayment cost.

The proof is also in Appendix.

4. Bayesian model

The purpose of this section is to obtain the conditional distribution of parameter θ for the distribution of prepayment cost, provided the history of riskless interest rate and prepayment events and to price the MBS in the setting that θ is uncertain in terms of the Bayesian model.

4.1 Description of posterior density

Remind us that $X(t)$ is a $(\mathcal{F}_t)$-progressively measurable process that may be interpreted as the MBS fair price when the parameter is unknown. From the result in the previous section, we have

$$X(t) = \int_{\mathbf{R}^d} X(t;\theta)Q(d\theta|\mathcal{F}_t), \tag{15}$$

where $Q(d\theta|\mathcal{F}_t)$ is the conditional distribution of parameter θ given $\mathcal{F}_t$.

If there exists a density $Q(\theta|\mathcal{F}_t)$, the Bayes formula implies that $Q(\theta|\mathcal{F}_t)$ can be calculated from the prior distribution $\psi(\theta)$ and the densities of $(r(s))_{0\leq s\leq t}$ and $(A(s))_{0\leq s\leq t}$ given by θ that generate the filtration $(\mathcal{F}_t)$ as is seen below.

It is unnecessary, however, to grasp the whole history of $(r(s))_{0\leq s\leq t}$ and $(A(s))_{0\leq s\leq t}$ for estimating θ. Necessary is only the prepayment costs observed and the upper bound of prepayment up to the time.

What we can observe at time t is

- the exact value of a prepayment cost if it is under the upper bound $U(t)$,
- only the fact that a prepayment cost must be above $U(t)$, otherwise.

The latter is treated as censored data in survival analysis.

Then we define the set of prepayment cost observed by time t by

$$\begin{aligned}
\beta(t) &\equiv \{b \in \{b_1, \cdots, b_N\} | b \leq U(t)\} \\
&= \{b \in \mathbf{R} | \tau(b) = s \text{ for } s \leq t\} \\
&= \{b \in \mathbf{R} | r(s) = S(s,b) \text{ and } \Delta A(s) = 1 \text{ for } s \leq t\}.
\end{aligned}$$

Let $D(t) = \{b_j > U(t); b_j \notin \beta(t)\}$ be the events of non prepayment up to time t. Besides, we suppose throughout this section that the distribution function $F(x;\theta)$ has a density function $f(x;\theta)$ under the probability measure Q.

Now we define the density of prepayment costs at time t as follows

$$\begin{aligned}
Q(\beta(t), D(t)|\theta) &= Q(\beta(t)|\theta, D(t))Q(D(t)|\theta) \\
&= \prod_{b_i \in \beta(t)} f(b_i;\theta) \times \{1 - F(U(t);\theta)\}^{N-A(t)}.
\end{aligned} \tag{16}$$

Then, from the Bayes formula and (16) it follows that

$$\begin{aligned} Q(\theta|\mathcal{F}_t) &= \frac{Q(\beta(t), D(t)|\theta)}{\int Q(\beta(t), D(t)|\theta)\psi(\theta)d\theta}\psi(\theta) \\ &= \frac{\prod_{b_i \in \beta(t)} f(b_i;\theta) \times \{1 - F(U(t);\theta)\}^{N-A(t)}\psi(\theta)}{\int \prod_{b_i \in \beta(t)} f(b_i;\theta) \times \{1 - F(U(t);\theta)\}^{N-A(t)}\psi(\theta)d\theta}. \end{aligned}$$

We can see that the effect of the path of the riskless interest rate remains only in terms of $U(t)$.

4.2 Example

The discussion of the last section implies that a likelihood which is often seen in survival analysis,

$$\prod_{b_i \in \beta(t)} f(b_i;\theta) \times \{1 - F(U(t);\theta)\}^{N-A(t)},$$

has an important role in evaluation of unknown parameter.

This section gives an illustration of using the actual data up to time t of riskless interest rate and prepayment events to calculate a maximum likelihood estimator $\hat{\theta}(t)$.

We present the implementation for a practical use through a simple example with an exponential distribution of prepayment cost.

Let $F(x;\theta)$ be an exponential distribution with parameter $\theta \geq 0$, and $f(x;\theta) = \theta e^{-\theta x}$ be its density.

We suppose $U(0) = 0$.

Now we assume that $\psi(\theta)$ is a normal distribution with mean $\theta_0 > 0$ and variance σ^2. It is probable that θ becomes negative in this case, but the possibility is almost negligible if σ is assumed to be small enough in comparison with θ_0. So we implicitly assume such a case.

Let the likelihood function with a prior density and its logarithm given by

$$\tilde{L}(\theta|\beta(t), D(t)) = \prod_{b_i \in \beta(t)} f(b_i;\theta) \times \{1 - F(U(t);\theta)\}^{N-A(t)}\psi(\theta)$$

and $\tilde{l}(\theta|\beta(t), D(t)) = \log \tilde{L}(\theta|\beta(t), D(t))$.

Hence we have the following likelihood equation:

$$\frac{\partial \tilde{l}}{\partial \theta} = \sum_{b_i \in \beta(t)} \left(\frac{1}{\theta} - b_i\right) - (N - A(t))U(t) - \frac{\theta - \theta_0}{\sigma^2} = 0.$$

At last, we obtain the maximum likelihood estimator up to time t, if $A(t) \geq 1$,

$$\begin{aligned}\hat{\theta}(t) &= \frac{\sigma^2}{2}\left(-\eta(t) + \sqrt{\eta(t)^2 + \frac{4}{\sigma^2}A(t)}\right)\\ &= \frac{A(t)}{\eta(t) + \sqrt{\eta(t)^2 + \frac{4}{\sigma^2}A(t)}}\end{aligned}$$

where

$$\eta(t) = \sum_{b_i \in \beta(t)} b_i + (N - A(t))U(t) - \frac{\theta_0}{\sigma^2}.$$

If $\beta(t) = \emptyset$, the above equation directly leads to

$$\hat{\theta}(t) = \theta_0 - N\sigma^2 U(t),$$

which become negative if $U(t) > \frac{\theta_0}{N\sigma^2}$.

Since the maximum likelihood estimator of θ should be non-negative, we take

$$\hat{\theta}(t) = \begin{cases} 0 & \text{if } U(t) > \frac{\theta_0}{N\sigma^2} \\ \theta_0 - N\sigma^2 U(t) & \text{if } U(t) \leq \frac{\theta_0}{N\sigma^2} \end{cases}$$

Next we present the calibration approach to estimate the parameters θ_0 and σ of the prior distribution.

Actual prepayment is announced every month in terms of Single Monthly Mortality (SMM), which is defined as the ratio of the prepayment (measured in terms of amounts of mortgage) during the last one month.

Since it is assumed that the mortgage size, that is, the initial principal is common to all the mortgagers, if SMM is announced at discrete times $t_0 < t_1 < \cdots < t_K$, SMM can be represented in terms of $A(t)$ as

$$SMM(t_k) = \frac{A(t_k) - A(t_{k-1})}{N - A(t_{k-1})}, \quad k = 1, \cdots, K.$$

If N is large enough, the law of large numbers implies that given $\mathcal{F}_{t_{k-1}}$, we see

$$
\begin{aligned}
& E^Q\Big[\frac{A(t_k) - A(t_{k-1})}{N - A(t_{k-1})}\Big|\mathcal{F}_{t_{k-1}}\Big] \\
= \quad & 1 - E^Q\Big[\frac{\frac{1}{N}\sum_{i=1}^N (1 - N(t_k, b_i))}{\frac{1}{N}\sum_{i=1}^N (1 - N(t_{k-1}, b_i))}\Big|\mathcal{F}_{t_{k-1}}\Big] \\
\approx \quad & 1 - \frac{1 - E^Q[N(t_k, b_1)|\mathcal{F}_{t_{k-1}}]}{1 - E^Q[N(t_{k-1}, b_1)|\mathcal{F}_{t_{k-1}}]} \\
= \quad & 1 - \frac{Q(\tau(b_1) > t_k|\mathcal{G}^r_{t_{k-1}} \vee \mathcal{H}^A_{t_{k-1}})}{Q(\tau(b_1) > t_{k-1}|\mathcal{G}_{t^r_{k-1}} \vee \mathcal{H}^A_{t_{k-1}})} \\
= \quad & \frac{E^Q[F(U(t_k); \hat{\theta}(t_{k-1}))|\mathcal{G}^r_{t_{k-1}}] - F(U(t_{k-1}); \hat{\theta}(t_{k-1}))}{1 - F(U(t_{k-1}); \hat{\theta}(t_{k-1}))}.
\end{aligned}
$$

The last equality follows from Lemma 3.1 and the indistinguishability of $b_1, \cdots, b_N$.

Hence we solve the following optimization problem so that θ_0 and σ can fit the actual available prepayment data although it is not so easy to solve the above problem.

$$
\text{minimize} \quad \sum_{k=1}^{K}\Big\{SMM(t_k) - \frac{E^Q[F(U(t_k); \hat{\theta}(t_{k-1}))|\mathcal{G}^r_{t_{k-1}}] - F(U(t_{k-1}); \hat{\theta}(t_{k-1}))}{1 - F(U(t_{k-1}); \hat{\theta}(t_{k-1}))}\Big\}^2.
$$

In this example, the optimization problem can be reduced to

$$
\text{minimize} \sum_{k=1}^{K}\{SMM(t_k) - (1 - E^Q[e^{-\hat{\theta}(t_{k-1})(U(t_k) - U(t_{k-1}))}|\mathcal{G}^r_{t_{k-1}}])\}^2,
$$

so it is necessary to achieve (at least approximate) the distribution of $U(t_k) - U(t_{k-1})$ to solve it.

This idea of calibration is almost the same as Kariya and Kobayashi [11], Nakamura [18] and so on.

5. Conclusion

In this article, we assume that each mortgager has an individual prepayment cost and reasonably judges to prepay the loan when the interest rate fell on a certain threshold dependent on her or his prepayment cost. Prepayment cost of each mortgager is specified as an independent random variable that follows a parametric probability distribution function. This means that the prepayment risk is formulated based on the idea that the investor in MBS cannot know the actual value until prepayment really occurs.

Under this situation, we first derived the theoretical price of pass-through type MBS under the assumption that the parameter is given. As a result, it is apparent how the burnout phenomenon and the current level of riskless interest rate have impacts upon the MBS price. Specifically, a sort of hazard function in the usage of default risk modeling appears in our model since prepayment cost is supposed to be directly unobservable.

In addition, we proposed the Bayesian model to estimate the parameter that decides the distribution of prepayment cost. We present a formula for the conditional density of the parameter under the condition that given are the exposed prepayment costs by actual events and the upper bound of prepayment cost of mortgagers who have prepaid. This leads to a maximum likelihood estimation by using the actual data, and we illustrate the estimation procedure in a simple case.

Some subjects are left as future research. We assume throughout this paper that each mortgager's prepayment cost is constant all the time although it is expected that prepayment costs may vary in time according to external state variables and each mortgager's reason.

Second, we see the model mostly from a viewpoint of investors who take part in the secondary market. If the actual minute prepayment data were available, we would study the relation between prepayment cost and other factors (age, income, region, and so on).

Anyhow, immaturity of MBS market in Japan has prevented us from collecting available data enough to test whether our model is advantage. Nevertheless, we strongly expect that public data like market price of MBS and prepayment frequency is more and more increasing so that the method proposed in this paper can be applied to the business in the future.

A. Appendix

A.1 Proof of Theorem 3.2

This section gives a proof to Theorem 3.2. The next proposition is essential to a series of our discussions.

Proposition A.1. *Define a stochastic process $M(t)$ by*

$$M(t) = A(t) - \int_0^t (N - A(s))d\Gamma(s).$$

Then $M(t)$ is a $(Q, (\mathcal{F}_t^\theta))$-martingale.

Proof. A new filtration $(\hat{\mathcal{F}}_t^\theta)$ is specified by

$$\hat{\mathcal{F}}_t^\theta = \bigcap_{s>t} \left(\mathcal{G}_s^{r,\theta} \vee \sigma\{N(u, b_i), i = 1, \cdots, N; u \le s\} \right).$$

Setting $L(t, b_i) = (1 - N(t, b_i))e^{\Gamma(t)}$, for $t \le s$, we have

$$\begin{aligned} E^Q[L(s, b_i)|\mathcal{G}_t^{r,\theta}] &= E^Q[e^{\Gamma(s)} E^Q[1 - N(s, b_i)|\mathcal{G}_s^{r,\theta}]|\mathcal{G}_t^{r,\theta}] \\ &= E^Q[e^{\Gamma(s)} e^{-\Gamma(s)}|\mathcal{G}_t^{r,\theta}] \\ &= 1. \end{aligned}$$

Hence

$$\begin{aligned} E^Q[L(s, b_i)|\hat{\mathcal{F}}_t^\theta] &= (1 - N(t, b_i)) \\ &\qquad \times \frac{E^Q[L(s, b_i)|\mathcal{G}_t^{r,\theta}, N(t, b_i) = 0]}{Q(N(t, b_i) = 0|\mathcal{G}_t^{r,\theta})} \\ &= (1 - N(t, b_i))e^{\Gamma(t)} E^Q[L(s, b_i)|\mathcal{G}_t^{r,\theta}] \\ &= (1 - N(t, b_i))e^{\Gamma(t)} \\ &= L(t, b_i). \end{aligned}$$

Therefore we observe $L(t, b_i)$ is a $(Q, (\hat{\mathcal{F}}_t^\theta))$-martingale.

Equivalently it follows

$$\begin{aligned} dL(t, b_i) &= -e^{\Gamma(t)} dN(t, b_i) \\ &\qquad + (1 - N(t, b_i))e^{\Gamma(t)} d\Gamma(t) \\ -e^{-\Gamma(t)} dL(t, b_i) &= dN(t, b_i) - (1 - N(t, b_i))d\Gamma(t). \end{aligned}$$

Thus defining letting $M(t, b_i)$ $i = 1, \cdots, N$ by

$$M(t, b_i) = N(t, b_i) - \int_0^t (1 - N(s, b_i))d\Gamma(s),$$

it follows $M(t, b_i)$ $i = 1, \cdots, N$ are $(Q, (\hat{\mathcal{F}}_t^\theta))$-martingales. Moreover

$$M(t) = \sum_{i=1}^N M(t, b_i) = A(t) - \int_0^t (N - A(s))d\Gamma(s)$$

also becomes a $(Q, (\hat{\mathcal{F}}_t^\theta))$-martingale.

Since $M(t)$ is $(\mathcal{F}_t^\theta)$-measurable and

$$\mathcal{H}_t^A \subset \sigma\{N(u, b_i), i = 1, \cdots, N; u \le t\}$$

(consequently $\mathcal{F}_t^\theta \subset \hat{\mathcal{F}}_t^\theta$), $M(t)$ follows a $(Q, (\mathcal{F}_t^\theta))$-martingale.

The last proposition leads to the following two consequences.

Proposition A.2. *For any $t \in [0, s]$ and any $\mathcal{G}_s^r$-measurable integrable random variable ξ, we have*

$$E^Q[Z(t,s)(N - A(s))\xi|\mathcal{F}_t^\theta] = (N - A(t)) \times E^Q[Z(t,s)e^{-(\Gamma(s)-\Gamma(t))}\xi|\mathcal{G}_t^{r,\theta}].$$

Proof. Let $L(t) = (N - A(t))e^{\Gamma(t)}$. By using (A.1), we notice

$$dL(t) = -e^{\Gamma(t)}dM(t).$$

This implies that $L(t)$ is a $(Q, (\mathcal{F}_t^\theta))$-local martingale.

Now define $Z(t)$ by

$$Z(t) = E^Q[Z(0,s)e^{-\Gamma(s)}\xi|\mathcal{G}_t^{r,\theta}].$$

We note that $Z(t)$ is a $(Q, (\mathcal{G}_t^{r,\theta}))$-continuous martingale, hence a $(Q, (\mathcal{F}_t^\theta))$-continuous martingale on $[0, s]$, due to the independence between $Z(0,s)e^{-\Gamma(s)}\xi$ and $\mathcal{H}_s^A$. Then the quadratic variation process $[Z, L](t) \equiv 0$. Therefore we have

$$d(Z(t)L(t)) = Z(t-)dL(t) + L(t)dZ(t),$$

and since $\Gamma(t)$ is $\mathcal{G}_t^{r,\theta}$-measurable,

$$\begin{aligned} |Z(t)L(t)| &= (N - A(t))E^Q[Z(0,s)e^{-(\Gamma(s)-\Gamma(t))}\xi|\mathcal{G}_t^{r,\theta}] \\ &\leq NE[|\xi|] < \infty. \end{aligned}$$

Hence $Z(t)L(t)$ follows a $(Q, (\mathcal{F}_t^\theta))$-martingale.

Since $Z(s)L(s) = Z(0,s)(N - A(s))\xi$, we obtain the desired equality as follows.

$$\begin{aligned} E^Q[Z(t,s)(N - A(s))\xi|\mathcal{F}_t^\theta] &= Z(0,t)^{-1}E^Q[Z(0,s)(N - A(s))\xi|\mathcal{F}_t^\theta] \\ &= Z(0,t)^{-1}E^Q[Z(s)L(s)|\mathcal{F}_t^\theta] \\ &= Z(0,t)^{-1}Z(t)L(t) \\ &= (N - A(t))E^Q[Z(t,s)e^{-(\Gamma(s)-\Gamma(t))}\xi|\mathcal{G}_t^{r,\theta}]. \end{aligned}$$

Proposition A.3.

$$\begin{aligned} &E^Q\left[\int_t^T Z(t,s)p(s)dA(s)\middle|\mathcal{F}_t^\theta\right] \\ &\quad = (N - A(t))E^Q\left[\int_t^T Z(t,s)p(s)e^{-(\Gamma(s)-\Gamma(t))}d\Gamma(s)\middle|\mathcal{G}_t^{r,\theta}\right] \end{aligned}$$

Proof. From Proposition (A.1), it follows

$$\int_t^T Z(t,s)p(s)dA(s) = \int_t^T Z(t,s)p(s)dM(s) - \int_t^T Z(t,s)p(s)(N-A(s))d\Gamma(s).$$

Since $M(t)$ is a $(Q, (\mathcal{F}_t^\theta))$-martingale,

$$Z(0,t)^{-1}E^Q\left[\int_t^T Z(0,s)p(s)dM(s)\middle|\mathcal{F}_t^\theta\right] = 0.$$

Hence we have

$$E^Q\left[\int_t^T Z(t,s)p(s)dA(s)\middle|\mathcal{F}_t^\theta\right] = E^Q\left[\int_t^T Z(t,s)p(s)(N-A(s))d\Gamma(s)\middle|\mathcal{F}_t^\theta\right].$$

Because $\Gamma(t)$ is non-decreasing in t, we can think of $\int_t^T Z(t,s)p(s)(N-A(s))d\Gamma(s)$ as Stieltjes integral. Therefore, for $t = t_0 < t_1 < \cdots < t_M = T$ with $\max_{1\le i\le M} |t_i - t_{i-1}| \to 0$ as $M \to \infty$, we can observe

$$\int_t^T Z(t,s)p(s)(N-A(s))d\Gamma(s) = \lim_{M\to\infty}\sum_{j=1}^M Z(t,t_{j-1})p(t_{j-1})(N-A(t_{j-1}))(\Gamma(t_j)-\Gamma(t_{j-1})).$$

Notice that we have

$$\sum_{j=1}^M Z(t,t_{j-1})p(t_{j-1})(N-A(t_{j-1}))(\Gamma(t_j)-\Gamma(t_{j-1})) \le Np(t)\Gamma(T)$$

and $E^Q[\Gamma(T)] < \infty$.

Thus it follows from Lebesgue's convergence theorem and Proposition A.2 that

$$
\begin{aligned}
& E^Q\left[\int_t^T Z(t,s)p(s)(N-A(s))d\Gamma(s)\middle|\mathcal{F}_t^\theta\right] \\
&= E^Q\left[\lim_{M\to\infty}\sum_{j=1}^{M} Z(t,t_{j-1})p(t_{j-1})(N-A(t_{j-1}))\right. \\
&\qquad\qquad \left.(\Gamma(t_j)-\Gamma(t_{j-1}))\middle|\mathcal{F}_t^\theta\right] \\
&= \lim_{M\to\infty}\sum_{j=1}^{M} E^Q\big[Z(t,t_{j-1})p(t_{j-1})(N-A(t_{j-1})) \\
&\qquad\qquad E^Q[\Gamma(t_j)-\Gamma(t_{j-1})|\mathcal{F}_{t_{j-1}}^\theta]|\mathcal{F}_t^\theta\big] \\
&= \lim_{M\to\infty}\sum_{j=1}^{M} E^Q\big[Z(t,t_{j-1})p(t_{j-1})(N-A(t_{j-1})) \\
&\qquad\qquad E^Q[\Gamma(t_j)-\Gamma(t_{j-1})|\mathcal{G}_{t_{j-1}}^{r,\theta}]|\mathcal{F}_t^\theta\big] \\
&= (N-A(t))\lim_{M\to\infty}\sum_{j=1}^{M} E^Q\big[Z(t,t_{j-1})p(t_{j-1}) \\
&\qquad\qquad e^{-(\Gamma(t_{j-1})-\Gamma(t))}E^Q[\Gamma(t_j)-\Gamma(t_{j-1})|\mathcal{G}_{t_{j-1}}^{r,\theta}]|\mathcal{F}_t^\theta\big] \\
&= (N-A(t))E^Q\left[\lim_{M\to\infty}\sum_{j=1}^{M} Z(t,t_{j-1})p(t_{j-1})\right. \\
&\qquad\qquad \left.e^{-(\Gamma(t_{j-1})-\Gamma(t))}(\Gamma(t_j)-\Gamma(t_{j-1}))\middle|\mathcal{F}_t^\theta\right] \\
&= (N-A(t))E^Q\left[\int_t^T Z(t,s)p(s)e^{-(\Gamma(s)-\Gamma(t))}d\Gamma(s)\middle|\mathcal{G}_t^{r,\theta}\right].
\end{aligned}
$$

In the last equality, we substitute $\mathcal{F}_t^\theta$ by $\mathcal{G}_t^{r,\theta}$ since the integral-form variable is independent of $A(t)$.

Applying the last two propositions to (10) leads to the desired result in Theorem 3.2.

A.2 Proof of Proposition 3.3

It follows from (13) and (14) that

$$dV(t) = V(t)\alpha(t)dB_t^Q - Z(0,t)e^{-\Gamma(t)}\big(c(t)dt + p(t)d\Gamma(t)\big).$$

Therefore we can see

$$\begin{aligned}
d\tilde{V}(t) &= d(Z(0,t)^{-1}V(t)) \\
&= r(t)Z(0,t)^{-1}V(t)dt + Z(0,t)^{-1}dV(t) \\
&= r(t)Z(0,t)^{-1}V(t)dt \\
&\quad + Z(0,t)^{-1}V(t)\alpha(t)dB_t^Q - e^{-\Gamma(t)}\big(c(t)dt + p(t)d\Gamma(t)\big) \\
&= \tilde{V}(t)\big(r(t)dt + \alpha(t)dB_t^Q\big) - e^{-\Gamma(t)}\big(c(t)dt + p(t)d\Gamma(t)\big).
\end{aligned}$$

This implies

$$\begin{aligned}
dX(t) &= d\big(N - A(t))e^{\Gamma(t)}\tilde{V}(t)\big) \\
&= -e^{\Gamma(t)}\tilde{V}(t-)dA(t) + (N - A(t))e^{\Gamma(t)}\tilde{V}(t)d\Gamma(t) \\
&\quad + (N - A(t))e^{\Gamma(t)}d\tilde{V}(t) \\
&= -(N - A(t-))^{-1}X(t-)dA(t) + X(t)d\Gamma(t) \\
&\quad + (N - A(t))e^{\Gamma(t)}\Big(\tilde{V}(t)\big(r(t)dt + \alpha(t)dB_t^Q\big) \\
&\quad - e^{-\Gamma(t)}\big(c(t)dt + p(t)d\Gamma(t)\big)\Big) \\
&= -(N - A(t-))^{-1}X(t-)dA(t) + X(t)d\Gamma(t) \\
&\quad + X(t)\big(r(t)dt + \alpha(t)dB_t^Q\big) \\
&\quad - (N - A(t))\big(c(t)dt + p(t)d\Gamma(t)\big).
\end{aligned}$$

Since, owing to (2),

$$\begin{aligned}
& dr(t) = \mu^Q(r(t),t)dt + \sigma(r(t),t)dB_t^Q \\
\Longleftrightarrow\quad & dB_t^Q = \sigma(r(t),t)^{-1}\big(dr(t) - \mu^Q(r(t),t)dt\big),
\end{aligned}$$

we obtain

$$\begin{aligned}
dX(t) &= \Big(X(t)\big(r(t) - \alpha(t)\sigma(r(t),t)^{-1}\mu^Q(r(t),t)\big) \\
&\quad - (N - A(t))c(t)\Big)dt \\
&\quad + X(t)\alpha(t)\sigma(r(t),t)^{-1}dr(t) \\
&\quad - (N - A(t-))^{-1}X(t-)dA(t) \\
&\quad + \big(X(t) - (N - A(t))p(t)\big)d\Gamma(t).
\end{aligned}$$

Furthermore Lemma 3.1 and (11) proves

$$d\Gamma(t) = \tilde{f}(U(t);\theta)dU(t), \tag{17}$$

where

$$\tilde{f}(U(t);\theta) = \frac{f(U(t);\theta)}{1 - F(U(t);\theta))}.$$

Now we achieve the desired result.

References

1. Deng, Y., Quingley, J.: Woodhead Behavior and the Pricing of Residential Mortgages. Berkeley Program on Housing and Urban Policy, Working Paper Series W00-004 (2001)
2. Duffie, D.: Dynamic Asset Pricing Theory 2nd edition. Princeton University Press 1996
3. Duffie, D., Lando, D.: Term structure of credit spreads with incomplete accounting information. Econometrica **69**, 633-664 (2001)
4. Dunn, K.B., McConnell, J.J.: A comparison of alternative models for pricing GNMA mortgage-backed securities. Journal of Finance **36**, 471-483 (1981a)
5. Dunn, K.B., McConnell, J.J.: Valuation of mortgage-backed securities. Journal of Finance **36**, 599-617 (1981b)
6. Dunn, K.B., Spatt, C.S.: The Effect of Refinancing Costs and Market Imperfections on Optimal Call Strategy and the Pricing of Debt Contracts. Working paper: Carnegie-Mellon University (1986)
7. Fabozzi, F.J.: The Handbook of Mortgage-Backed Securities. McGraw-Hill 1995
8. Ichijo, H., Moridaira, S.: Analysis on Prepayment of Mortgage Loans. The 15th JAFEE Conference Proceedings (Japanese) 2001
9. Jeanblanc, M., Rutkowski, M.: Modelling of Default Risk: An Overview. Working paper (1999)
10. Kagraoka, Y.: OAS Approach and Martingale Measure for Mortgage Prepayment. Working Paper (2001)
11. Kariya, T., Kobayashi, M.: Pricing mortgage-backed securities (MBS). Asia-Pacific Financial Markets **7**, 189-204 (2000)
12. Kariya, T., Uchiyama, F., Pliska, S.R.: A 3-factor Valuation Model for Mortgage-Backed Securities (MBS). Working paper (2002).
13. Kusuoka, S.: A remark on default risk models. Advances in Mathematical Economics **1**, 69-82 (1999)
14. Lamberton, D., Lapayre, B.: Introduction au Calcul Stochastique Appliqué à la Finance, 2nd. ed. Ellipses 1997
15. Liptser, R.S., Shiryayev, A.N.: Statistics of Random Processes I. General Theory. 2nd edition. Applications of Mathematics **5**. Springer-Verlag 2001
16. Mcconnell, J.J., Singh, M.: Rational prepayment and the valuation of collateralized mortgage obligations. Journal of Finance **49**, 891-921 (1994)
17. Nakagawa, H.: A Filtering model on default risk. Journal of Mathematical Sciences: University of Tokyo **8**, 107-142 (2001)
18. Nakamura, N.: Valuation of Mortgage-Backed Securities Based upon a Structural Approach. Working paper (2001)
19. Nakamura, N., Shibasaki, K.: Valuation of Mortgage-Backed Securities Based upon a Hazard Rate Approach. The 15th JAFEE Conference Proceedings (Japanese) 2001
20. Richard, S.F., Roll, R.: Prepayments on fixed-rate mortgage-backed securities. Journal of Portfolio Management **15**, 73-82 (1989)
21. Schwartz, E.S., Torous, W.N.: Prepayment and the valuation of mortgage-backed securities. Journal of Finance **44**, 375-392 (1989)
22. Stanton, R.: Rational prepayment and the valuation of mortgage-backed securities. Review of Financial Studies **8**, 677-708 (1995)

Adv. Math. Econ. 6, 149–165 (2004)

Advances in
MATHEMATICAL
ECONOMICS

Fixed point theorems in Hausdorff topological vector spaces and economic equilibrium theory

Ken Urai[1]* and Akihiko Yoshimachi[2]

[1] Faculty of Economics, Osaka University, Machikaneyama, Toyonaka, Osaka 560-0043, Japan
(e-mail: urai@econ.osaka-u.ac.jp)

[2] Graduate School of Economics, Osaka University, Machikaneyama, Toyonaka, Osaka 560-0043, Japan

Received: July 22, 2003
Revised: October 9, 2003

JEL classification: C62, D50

Mathematics Subject Classification (2000): 47H10, 91B50

Abstract. The aim of this paper is to develop fixed point theorems in Hausdorff topological vector spaces that are suitable for the purpose of economic equilibrium theory. The special concept we have used here is the "direction structure" that characterizes mappings in the economic theory, (preferences, excess demands, and the like,) adequately, and enables us to modify problems on mappings into those on a structure of the base set. Especially, since our mathematical generalization may directly be related to the continuity and/or convexity of individual preferences, we may obtain existence theorems of maximal points, Pareto optimal allocations, and price equilibria for Gale-Nikaido-Debreu abstract economies under quite natural conditions.

Key words: fixed point theorem, non-ordered preference, direction structure, Gale-Nikaido-Debreu theorem, market equilibrium

1. Introduction

In the economic theory, fixed point theorem is one of the most important mathematical tools that enable us to construct the concept of economic equilibria. Under the earliest formulation as in Arrow and Debreu (1954), Nikaido (1956), etc., the economic equilibrium was treated as a fixed point

* Our special thanks are due to Professor Toru Maruyama (Keio University) and an anonymous referee.

of a continuous mapping constructed by continuous excess demands and price formation functions. The problem was lately reformulated as a general coincidence property for restriction and preference correspondences including cases with non-ordered preferences (c.f. Shafer and Sonnenschein (1975), Gale and Mas-Colell (1975)).

In this paper, we prove fixed point theorems and theorems on economic equilibria under weak conditions on local directions of mappings in Hausdorff topological vector spaces. Our generalization directly aims to support a weak condition on the convexity and continuity of preference correspondences for the existence of economic equilibria.

Results in this paper are based on recent researches in Urai (2000), Urai and Hayashi (2000), and Urai and Yoshimachi (2002) on fixed point theorems for multi-valued mappings and economic equilibria in Hausdorff topological vector spaces. We have developed here a way to generalize them by using more essential mathematical structure defining a notion of "directions" in topological vector spaces. The notion enables us to characterize mappings in economics (especially, preference mappings,) adequately, and to modify problems on properties of such mappings into problems on properties of spaces. Continuity for the direction of set valued mappings may be reduced into topological features for the direction structure of spaces, and the generalization of continuity condition for the existence of fixed points as in Urai (2000) may be reformulated here as a generalization of conditions for a subset of vector spaces on which all continuous functions have fixed points.

In this paper, we apply these results to general existence theorems of maximal points, Pareto optimal allocations, and price equilibria for an abstract economy of the Gale-Nikaido-Debreu type. The result may also be applied to the existence of equilibrium for an abstract economy of the Arrow-Debreu type with (possibly) non-ordered, non-convex, and/or non-continuous preferences and constraint correspondences in Hausdorff topological vector spaces which may not necessarily be locally convex[1].

2. Fixed point theorems

In this section, we define a structure on a topological vector space which generalizes some concepts included in our ordinary usage of the word a "direction" for a vector, and we show fixed point theorems depending merely on local conditions for such directions associated with mappings without using the concept of continuity and/or convexity. Moreover, the structure is also used to describe

[1] See also Urai-Yoshimachi(2002) for a similar result based on Urai-Hayashi (2000), where a direction of mapping is treated as an element of the dual space of a locally convex base space.

a weaker condition on a subset of vector spaces under which every continuous function has a fixed point.

Let E be a Hausdorff topological vector space over the real field R, and let X be a subset of E. We define a structure which represents the set of points in *direction* $y - x$ *at* x *in* X for each $x, y \in X$ as follows. For each pair (x, y) of an element of $X \times X$, define a subset $V(x, y)$ of X satisfying:

(A0) $\forall x, y \in X$, $V(x, y)$ is a convex subset of X.
(A1) $\forall x, y \in X$, $x \notin V(x, y)$,
(A2) $\forall x, y, z \in X$, $(z \in V(x, y)) \to (y \in V(x, z))$ [2].

We call the set $V(x, y) \subset X$ the *set of points in direction* $y - x$ *at* x, or simply, the *set of direction* y *from* x, and we say that X has a *direction structure* V [3].

An example for such a structure is obtained by using the inner product. When the inner product is defined on E, we may define *the inner product direction structure*, V on $X \subset E$, as $V(x, y) = \{z \in X | \langle y - x, z - x\rangle > 0\}$[4]. For a vector space E with an algebraic dual E^*, we may also define V on $X \subset E$ as $V(x, y) = \{z \in X | p(x, y)(z - x) > 0 \wedge p(x, z)(y - x) > 0\}$ whenever $p : X \times X \to E^*$ is a mapping such that $p(x, x) = 0$ for all $x \in X$. Another example may be obtained when there is a correspondence $\varphi : X \to X$ such that $x \notin \operatorname{co} \varphi(x)$ for all $x \in X \setminus K$, where $K = \{x \in X | x \in \varphi(x)\}$ is the *fixed point set* of φ [5]. In this case, we may define direction structure V_φ on $X \subset E$ as $V_\varphi(x, y) = X \cap \operatorname{co} \varphi(x)$, if $x \in X$ and $y \in \operatorname{co} \varphi(x)$, else $V_\varphi(x, y) = \emptyset$. We call direction structure V_φ the *direction structure induced by* φ on X.

Suppose that X has a direction structure, $V : X \times X \to X$. We say that a correspondence (possibly empty valued) $\varphi : X \to X$ *has a locally common direction* y^x *at* x (under V) if there exists an open neighbourhood $U(x)$ of x such that $\varphi(z) \subset V(z, y^x)$ for all $z \in U(x)$. Based on the direction structure, we have the following fixed point theorem.

Theorem 1. *(Fixed Point Theorem for Mappings Having Locally Common Directions) Let X be a non-empty compact convex subset of Hausdorff topological vector space E having direction structure V, and let $\varphi : X \to X$ be a non-empty valued correspondence. Suppose that φ has a locally common direction under V at every x such that $x \notin \varphi(x)$. Then, φ has a fixed point.*

---

[2] Condition (A2) may be replaced with a weaker condition (A2') $\forall x, y, z \in X$, $(w \in V(x, y) \cap V(x, z)) \to (\forall \lambda \in [0, 1], w \in V(x, \lambda y + (1 - \lambda) z))$. Note that under (A1) and (A2), $V(x, x) = \emptyset$ for all $x \in X$.

[3] The similar notion has already been treated in Urai (2000) [Concluding Remarks].

[4] Throughout this paper, we denote by $\langle \cdot, \cdot \rangle$ the vector space duality operation including the inner product.

[5] The notation $\operatorname{co} A$ denotes the convex hull of A.

Proof. Assume that φ has no fixed point. Then, φ has a locally common direction at each $x \in X$. Since X is compact, we have points $x^1, \cdots, x^n \in X$, open neighbourhoods $U(x^1), \cdots, U(x^n)$ of each $x^1, \cdots, x^n$ in X such that $\bigcup_{t=1}^n U(x^t) \supset X$, together with points $y^{x^1}, \cdots, y^{x^n} \in X$ satisfying for each $t = 1, \cdots, n$, $\varphi(z) \subset V(z, y^{x^t})$ for all $z \in U(x^t)$. Let $\beta_t : X \to [0,1]$, $t = 1, \cdots, n$, be a partition of unity subordinated to $U(x^1), \cdots, U(x^n)$. Let us define a function f on $D = \mathrm{co}\,\{y^{x^1}, \cdots, y^{x^n}\}$ to itself as $f(x) = \sum_{t=1}^n \beta_t(x) y^{x^t}$. Then, f is a continuous function on the finite dimensional compact set D to itself. Hence, f has a fixed point $z^* = f(z^*) = \sum_{t=1}^n \beta_t(z^*) y^{x^t}$ by Brouwer's fixed point theorem. On the other hand, for all t such that $z^* \in U(x^t)$, (i.e., $\beta_t(z^*) > 0$,) $\varphi(z^*) \subset V(z^*, y^{x^t})$, so that (by condition (A2)) for an element $y^* \in \varphi(z^*)$ arbitrarily fixed, we have $y^{x^t} \in V(z^*, y^*)$ for all t such that $\beta_t(z^*) > 0$. Since $V(z^*, y^*)$ is convex, we have $z^* = \sum_{t=1}^n \beta_t(z^*) y^{x^t} \in V(z^*, y^*)$, which contradicts the fact $z^* \notin V(z^*, y^*)$ under condition (A1). □

Every non-empty convex valued correspondence having open lower sections, $\varphi : X \to X$, has a locally common direction at each x such that $x \notin \varphi(x)$ under the direction structure induced by φ. Hence, by considering the induced direction structure, the above theorem may be considered as an extension of Browder's fixed point theorem (c.f., Browder (1968)). The theorem also includes one of the main fixed point theorems in Urai (2000) (Theorem 1 (K^*)). In the following, by using the concept of direction structure, we further extend the result to more general cases with mappings having locally "continuous" directions[6].

We say that a direction structure, $V : X \times X \to X$, is *lower topological* on a certain subset $A \subset X \times X$ if the following (A3) is satisfied [7].

> (A3) For all $(x, y) \in A$, $V(x, y) \neq \emptyset$ implies that $\exists W(x, y)$, an open neighbourhood of $(x, y) \in A \subset X \times X$ such that $\bigcap_{(z,w) \in W(x,y)} V(z, w) \neq \emptyset$.

A correspondence $\varphi : X \to X$ is said to *have a locally continuous direction at* x (under a structure V) if there exists an open neighbourhood $U(x)$ and a continuous function $y : U(x) \to X$ such that $\varphi(z) \subset V(z, y(z))$ for all $z \in U(x)$.

---

[6] The difference is important especially when values of mappings are closed, e.g., if $f : X \to X$ is single valued, under the induced direction structure, f fails to have a locally common direction as long as it is not locally constant, though f always has a continuous direction as long as it is continuous.

[7] Indeed, the following is nothing but a condition for the lower section of the correspondence V, i.e., for each $V(x, y) \neq \emptyset$ there is at least one element whose lower section is a neighbourhood of (x, y).

Theorem 2. *(Fixed Point Theorem for Mappings Having Locally Continuous Directions I) Let X be a non-empty compact convex subset of Hausdorff topological vector space E having lower topological direction structure V on $X \times X$, and let $\varphi : X \to X$ be a non-empty valued correspondence. Suppose that φ has a locally continuous direction (under V) at every x such that $x \notin \varphi(x)$. Then, φ has a fixed point.*

Proof. Assume that φ has no fixed point. Then, φ has a locally continuous direction $y^x : U(x) \to X$, $\varphi(z) \subset V(z, y^x(z))$ for all $z \in U(x)$ at each $x \in X$. Since X is compact, we have finite points $x^1, \cdots, x^n \in X$, open neighbourhoods $U(x^1), \cdots, U(x^n)$ of each $x^1, \cdots, x^n$ in X such that $\bigcup_{t=1}^n U(x^t) \supset X$, together with continuous functions $y^{x^1} : U(x) \to X, \cdots, y^{x^n} : U(x) \to X$ satisfying for each $t = 1, \cdots, n$, $\varphi(z) \subset V(z, y^{x^t}(z))$ for all $z \in U(x^t)$. Let $\beta_t : X \to [0,1]$, $t = 1, \cdots, n$, be a partition of unity subordinated to $U(x^1), \cdots, U(x^n)$. Let us define a function f on X to itself as $f(x) = \sum_{t=1}^n \beta_t(x) y^{x^t}(x)$, where $y^{x^t}(x)$ denotes 0 for each $x \notin U(x^t)$. Then, f is a continuous function on X to itself. Since $\emptyset \neq \varphi(z) \subset V(z, \sum_{t=1}^n \beta_t(z) y^{x^t}(z))$ for all $z \in X$, by defining $\Phi(z)$ as $\Phi(z) = V(z, f(z))$, the mapping $\Phi : X \to X$ is non-empty valued and $z \notin \Phi(z)$ for all $z \in X$. Moreover, by (A3) and by the continuity of f, Φ has a locally common direction at each $x \in X$. Hence, by Theorem 1, Φ has a fixed point, so that we have a contradiction. □

When the topology on a vector space is given by the inner product, the inner product direction structure, $V(x,y) = \{z | \langle z - x, y - x \rangle > 0\}$, is lower topological. Hence, the theorem gives a sufficient generality for finite dimensional cases under the inner product direction structure. The following corollary which is an immediate consequence of the above argument shows one of the most interesting (and important) features of our theorem. Note that no continuity and no convexity are assumed on mapping P except for those on the function, d [8].

Corollary 2.1. *Let X be a non-empty compact convex subset of R^n and let $P : X \to X$ be a correspondence having continuous direction $d : X \to X$ under the Euclidean inner product direction structure, i.e., $D(x) \cdot (y - x) > 0$ for all $y \in P(x)$, where $D(x) = d(x) - x$ for each $x \in X$. Then, there is a point x such that $P(x) = \emptyset$* [9].

Proof. Note that P has no fixed point since $D(x) \cdot (y - x) > 0$ for all $y \in P(x)$ for all $x \in X$. If P is non-empty valued, however, under the inner product direction structure, P has a fixed point x^* by Theorem 3. □

[8] For the Euclidean inner product in R^n, we write $x \cdot y$ instead of $\langle x, y \rangle$.

[9] In view of economics, above $D(x)$ may be interpreted as a generalized concept for the continuous first derivative of a utility function.

A slight modification of the above theorem into the case with single valued mapping will be useful in the later. (Note that the next corollary is trivial when the topology on space E is locally convex by the fixed point theorem of Schauder-Tychonoff.)

Corollary 2.2. *(Fixed Point Theorem for a Continuous Mapping I) Let X be a non-empty compact convex subset of Hausdorff topological vector space E and let $f : X \to X$ be a continuous function. Suppose that there is a direction structure V on X such that $f(x) \in V(x, f(x))$ at each $x \neq f(x)$ and V is lower topological on the graph of f. Then, f has a fixed point.*

Proof. Assume the contrary. Then, the mapping $\Phi(x) = V(x, f(x))$ on X to X is non-empty convex valued. Moreover, Φ has a locally common direction at each $x \in X$ since f is continuous and (A3) is satisfied at each $(x, f(x))$. Hence, Φ has a fixed point by Theorem 2, so that we have a contradiction. □

By reading the proof of Theorem 3 or Corollary 3.5, one can see that the lower topological property, (A3), is nothing but a sufficient condition for cases with locally "continuous" directions to be reduced to cases with locally "common" directions in Theorem 2. Unfortunately, however, except for the above inner product cases, there seems to be no obvious way to define a direction structure which is lower topological on $X \times X$. Even for cases such that E together with the topological dual E' forms the duality, and $V(x, y) = \{z \in X | \langle p(x, y), z - x \rangle > 0\}$, where p is a function on $X \times X$ to E', defines a direction structure, we should induce a compact convergence topology on E' since for V to be lower topological, we have to assure the duality operation, $\langle \cdot, \cdot \rangle$ to be jointly continuous [10].

For fixed point arguments in more general topological spaces, it is more desirable to use the following alternative condition, (A4), to (A3) [11]. We say that a direction structure, $V : X \times X \to X$, is *upper topological* on a certain subset $A \subset X \times X$ if the following (A4) is satisfied [12].

(A4) For all $(x, y) \in A$, if $V(x, y) \neq \emptyset$, then there are two neighbourhoods U_x of x and U'_y of y in X such that $U_x \cap \operatorname{co} U'_y = \emptyset$.

See Figure 1. We are considering the case that the space E may not be locally convex, so that there may not exist a convex neighbourhood base at each point.

[10] Under the compact convergence topology on E', however, we further generalize our results to the case with mappings having *compact valued upper semicontinuous directions*. See, Urai-Yoshimachi (2002).

[11] Of course, it is not saying that condition (A4) is more general than condition (A3). Indeed, in the proof of next theorem, (A4) is used as a sufficient condition for the space to have a certain lower topological direction structure.

[12] As stated below, the condition is closely related to a property for the upper section of the correspondence, V.

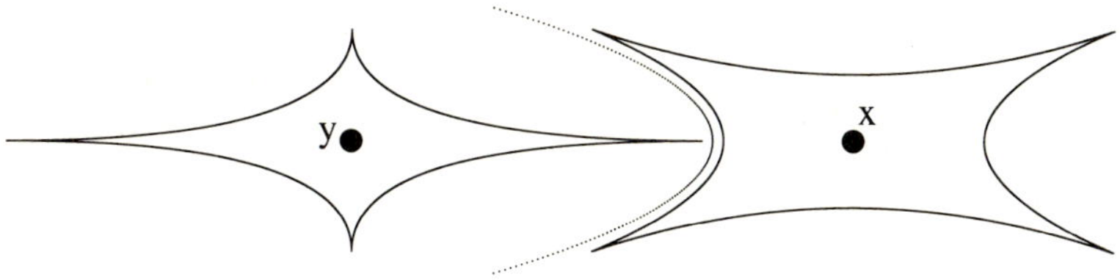

Fig. 1. Condition (A4)

Condition (A4) has its meaning only for cases with $V(x,y) \neq \emptyset$. If we do not allow $V(x,y)$ to have empty value as long as $x \neq y$, then a typical condition assuring (A4) is that for each x, y, $x \neq y$, $V(x,y)$ includes at least one neighbourhood of y. Hence, (A4) is typically a condition for the upper section of the correspondence, V [13]. Now we have the following theorem.

Theorem 3. *(Fixed Point Theorem for Mappings Having Locally Continuous Directions II) Let X be a non-empty compact convex subset of Hausdorff topological vector space E having upper topological direction structure V on $X \times X$, and let $\varphi : X \to X$ be a non-empty valued correspondence. Suppose that φ has a locally continuous direction (under V) at every x such that $x \notin \varphi(x)$. Then, φ has a fixed point.*

Proof. As in the proof of Theorem 3, assume that φ has no fixed point, and obtain the continuous function $f : X \to X$. Note that the non-emptiness of $V(x, f(x))$ for all $x \in X$ means that f has no fixed point. Let $A \subset X \times X$ be the graph of f. On A, we modify the direction structure V into V' as follows. For each $(x, f(x)) \in A$, $V(x, f(x)) \neq \emptyset$ implies (under (A4)) that there are two disjoint neighbourhoods, U_x and $U'_{f(x)}$, of x and $f(x)$ such that $U_x \cap \operatorname{co} U'_{f(x)} = \emptyset$. Hence, by the continuity of f, there is an open neighbourhood $O_x \subset U_x$ of x such that for all $z \in O_x$, $f(z) \in U'_{f(x)}$ and $O_x \cap \operatorname{co} U'_{f(x)} = \emptyset$. Moreover, since for a vector space topology it is always possible to take a closed neighbourhood base, O_x may be chosen so that there are two neighbourhoods of $f(x)$ in X, $U^1_{f(x)}$, $C^1_{f(x)}$, satisfying:

(1) $\operatorname{int} U'_{f(x)} \supset U^1_{f(x)} \supset C^1_{f(x)}$ [14],

[13] In such a case, by considering $V(z,x)$ and $V(z,y)$ for $z = (x+y)/2$, we have two disjoint convex neighbourhoods of x and y for each $x \neq y$. Note, however, that the whole space, E, may not be locally convex even in such cases. For an example, consider ℓ^p for $0 < p < 1$.

[14] Notation int A denotes the interior of A. In this proof, all interiors are taken with respect to the topology on the whole space, E.

(2) $U^1_{f(x)}$ is open and $C^1_{f(x)}$ is closed (i.e., compact) in X,
(3) $\forall z \in O_x$, $z + (U'_{f(x)} - x) \supset U^1_{f(x)}$,
(4) $\forall z \in O_x$, $f(z) \in C^1_{f(x)}$,
(5) O_x is closed in X.

Indeed, by the property of vector space topology, we may chose a circled 0 neighbourhood U in E so that $(U + U) \cap X \subset U'_{f(x)} - f(x)$ [15]. Then, by taking O_x so small that $O_x \subset x + U$, we have for all $z \in O_x \subset x + U$, $((x - z) + U) \cap X \subset (U + U) \cap X \subset U'_{f(x)} - f(x)$, i.e., $(f(x) + U) \cap X \subset z + (U'_{f(x)} - x)$. Hence, by setting $U^1_{f(x)} = (f(x) + \operatorname{int} U) \cap X$ and choosing $O_x \subset (x + \operatorname{int} U) \cap X$, (1),(2),(3) are satisfied. For conditions (4) and (5), chose $C^1_{f(x)} \subset U^1_{f(s)}$ as an arbitrary closed neighbourhood of $f(x)$ and redefine a closed O_x (smaller than before) by considering the continuity of f. On each O_x, define $V^x(z, f(z))$ as $(z + (\operatorname{int}(\operatorname{co} U'_{f(x)}) - x)) \cap X \supset U^1_{f(x)} \supset C^1_{f(x)} \supset f(O_x)$. Since X is compact, there are finite points $x^1, \ldots, x^m$ such that $O_{x^1}, \ldots, O_{x^m}$ covers X. For each $(x, f(x)) \in A$, let

$$V'(x, f(x)) = \bigcap_{t \in I(x)} V^{x^t}(x, f(x)),$$

where $I(x) \neq \emptyset$ denotes the subset of $\{1, 2, \ldots, m\}$ such that $(t \in I(x)) \iff (x \in O_{x^t})$. $V'(x, f(x))$ is non-empty since $(t \in I(x)) \iff (x \in O_{x^t}) \Longrightarrow (f(x) \in C^1_{f(x^t)})$. Moreover, by defining $V'(x, y) = \emptyset$ for all $y \notin V'(x, f(x))$, V' is a direction structure which is lower topological on $X \times X$. Indeed, for each $(z, f(z))$, let I_z be the set of index t such that $z \in O_{x^t}$. Then, by defining W^1 as a neighbourhood of z in X such that W^1 does not intersect O_{x^t} for all $t \notin I_z$, and W^2 as $\bigcap_{t \in I_z} U^1_{f(x^t)}$, W^1 and W^2 are open neighbourhoods of z and $f(z)$, respectively, and for $W = W^1 \times W^2$, $\bigcap_{(x,y) \in W} V'(x, y) \ni f(z)$. This, together with the openness for the value of V', implies that V' satisfies condition (A3). Hence, f has a fixed point by Corollary 3.5. Since f has no fixed point, we have a contradiction. □

An important corollary to the above theorem is the case that the mapping φ is a continuous function $f : X \to X$. In this case, we are able to rewrite the condition with more familiar concepts without an essential loss in generality.

Corollary 3.1. *(Fixed Point Theorem for a Continuous Mapping II) Let X be a non-empty compact convex subset of Hausdorff topological vector space E such that*

> *(T) every two different points, $x \neq y$, in E has two disjointed neighbourhoods, $U_x \cap U_y = \emptyset$, at least one of which is convex.*

[15] See Schaefer (1971) [p.14, Theorem 1.2].

Then, every continuous function, $f : X \to X$, has a fixed point.

Proof. Note that in this proof, condition (T) is used merely for pairs of points on the graph of f, i.e., x and $f(x)$ such that $x \neq f(x)$. Assume that f has no fixed point. Then, for each $(x, f(x)) \in X \times X$, x and $f(x)$, respectively, have neighbourhoods, U_x and $U_{f(x)}$, such that $U_x \cap U_{f(x)} = \emptyset$ and at least one of which is convex. Since the topology of vector space is translation invariant, we may suppose $U_{f(x)}$ is convex without loss of generality. Hence, by defining $V(x, f(x)) = U_{f(x)}$ for each $x \in X$ and $V(x, y) = \emptyset$ for each x and $y \notin U_{f(x)}$, we obtain a direction structure which is upper topological on $X \times X$. Since f has a continuous direction under V at everywhere, f has a fixed point by Theorem 6, so that we have a contradiction. □

The above condition, (T), is automatically satisfied when the topology on E is locally convex. The converse is not true, i.e., there is a topological vector space whose topology is not locally convex but satisfies condition (T) [16]. Hence, Corollary 6.7 is an extension of Schauder-Tychonoff's fixed point theorem. Since our main purpose is a generalization of the continuity condition for set valued mappings, it seems that there is no advantage in our approach for cases with continuous functions. The above corollary suggests, however, that our characterization of mappings under the direction structure seems to exhaust all the essential features in the notion of continuity at least for fixed point arguments.

As an important result for cases with set valued mappings, we show the following corollary. Also in this result, we reformulate our concept into more familiar notion, the existence of (local) continuous selections.

Corollary 3.2. *(Fixed Point Theorem for Mappings Having Continuous Local Selections) Let X be a non-empty compact convex subset of Hausdorff topological vector space E satisfying condition (T), and let $\varphi : X \to X$ be a non-empty convex valued correspondence. If φ has, locally, a continuous selection at each $x \in X$ such that $x \notin \varphi(x)$, then φ has a fixed point.*

Proof. Suppose that φ has no fixed point. Then, for each $x \in X$, there is an open neighbourhood U_x of x and a continuous function $f^x : U_x \to X$ such that $f^x(z) \in \varphi(z)$ for all $z \in U_x$. Since X is compact there are finite covering $U_{x^1}, \ldots, U_{x^m}$ and local selections, $f^{x^1}, \ldots, f^{x^m}$. Let $\alpha^t : U_{x^t} \to [0, 1]$, $t = 1, \ldots, m$ be the partition of unity subordinated to the finite covering. Then, $f : X \to X$ defined as $f(x) = \sum_{t=1}^m \alpha^t(x) f^{x^t}(x)$, where $f^{x^t}(x)$ for $x \notin U_{x^t}$ is defined as 0, is a continuous selection of φ since φ is convex valued. Then, by Corollary 6.7, f has a fixed point, so that φ has a fixed point, a contradiction. □

[16] A simple example is the space ℓ^p for $0 < p < 1$ under the pseudo ℓ^p-norm, $\|(x_i)_{i=1}^\infty\| = \sum_{i=1}^\infty |x_i|^p$.

In the above proof, if φ has a special type of local continuous selections, local selections of constant functions, (e.g., if φ has an open lower section at everywhere), then condition (T) will be omitted. Indeed, for such a case, the range of f is finite dimensional, so that Brouwer's fixed point theorem is sufficient for f to have a fixed point. The situation is completely the same when the range of each local continuous selection is not a single point (constant function) but a subset of finite dimensional space. This may also be considered as a corollary to the case that we have treated in Theorem 2 as the locally common direction. But we shall write it here as the most simple case of the previous corollary.

Corollary 3.3. *(A Generalization of Browder's Theorem) Let X be a non-empty compact convex subset of Hausdorff topological vector space E, and let $\varphi : X \to X$ be a non-empty convex valued correspondence. If φ has, locally, a continuous selection whose range is a subset of finite dimensional subspace at each $x \in X$ such that $x \notin \varphi(x)$. Then, φ has a fixed point.*

Proof. As stated above, since the range of each f^{x^t}, $f^{x^t}(U_{x^t})$ in the proof of the previous corollary is in a finite dimensional subspace L_t of E, the range of f is also in a finite dimensional subspace $L = \sum_{t=1}^{m} L_t$ of E. Hence, the function, f, restricted on $L \cap X$ has a fixed point by Brouwer's fixed point theorem, so that we have a contradiction. □

These corollaries shows that there may exist a trade-off between the generality for the vector space topology on the base set and the variety of mappings to which we want to show the existence of fixed points. Our approach, however, also suggests that the concept of direction structure brings about a unified view point on these topologies and mappings. For example, compare Corollary 3.5 and Corollary 6.7. In Corollary 6.7, the condition for the direction structure is completely described as a property on the topological vector space, Condition (T). On the contrary, in Corollary 6.7, the condition is described, completely, as a property for the mapping f.

3. Theorems on economic equilibria

In this section, we apply fixed point theorems in the previous section to several problems in the economic equilibrium theory.

Let $P : X \to X$ be a (possibly empty valued) correspondence on a subset X of a topological vector space E to itself. Assume that P satisfies

(Irreflexivity) $\forall x \in X, x \notin P(x)$.

In the following, we regard X as an individual choice set and $P(x) \subset X$ as the set of points which are preferred to x for each $x \in X$. Then, an element $x^* \in X$

may be interpreted as a *maximal element* for the preference correspondence, P, if $P(x^*) = \emptyset$.

In the previous section, we have seen in Corollary 3.4 that a fixed point theorem may easily be modified to the existence theorem on maximal elements. I.e., the existence of maximal elements for an irreflexive mapping may be considered as a contrapositive assertion to the existence of fixed point for a non-empty valued correspondence. Since in the maximal element existence problem, mapping P directly represents the individual preferences, the importance of our generalization of mappings for fixed point theorem (in the previous section) should be measured by the generality for a representation of our general preferences.

We emphasize that as a condition for the preference, "better sets have locally similar directions" is not only mathematically general but also intuitively natural. It is far more natural than the continuity. Moreover, there are many concrete examples that may not be treated in the standard argument but may be treated in our scope. For example, there is an important sort of *ordered* preferences that are *not continuous* to fail to have open lower sections, the *lexicographic ordering*. There also exists an important sort of ordered and continuous preferences that are *not complete* to fail to have open lower sections (consider the relation $\leq$ in R^n representing the strict monotonicity at each point). (See Figure 2. In each case, the better set at x is denoted by the shaded area.)

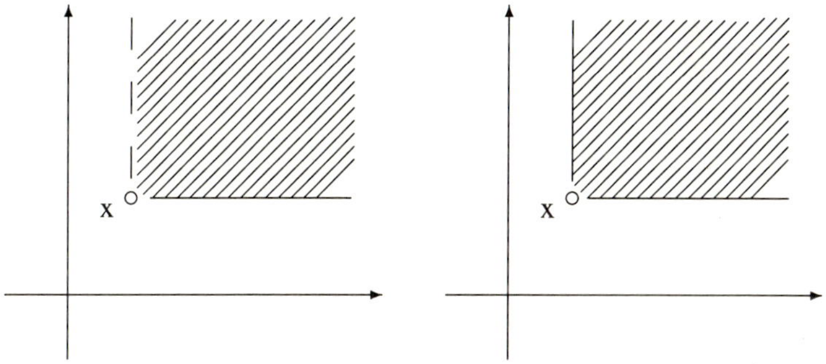

Fig. 2. Lexicographic orderings and orderings in vector spaces

We write here a generalized version of Corollary 3.4, a theorem on the existence of maximal elements, whose assumptions are taken as weak as possible from the economic view point. Preceding the theorem, we describe several basic settings for the economic equilibrium arguments.

(B0) [Basic Settings for the Economic Equilibrium Theory]: Let E be a Hausdorff topological vector space, X be a compact subset of E, and E'_X be a convex subset of the set of linear functionals on E, E^*, having a Hausdorff vector space topology such that the closure D of E'_X in E^* is compact, and the restriction of bilinear form $\langle p, x \rangle$ on $E'_X \times X$ is jointly continuous [17].

We shall use above basic settings, repeatedly, throughout this section.

Theorem 4. *(Existence of Maximal Elements) Under (B0), suppose that $P : X \to X$ is a (possibly empty valued) correspondence such that for each $x \in X$, there is a neighbourhood U_x of x and an upper semicontinuous compact valued correspondence $\theta_x : U_x \to E'_X$ satisfying that $\langle p, w - z \rangle > 0$ for all $p \in \theta(z)$, $w \in P(z)$, and $z \in U_x$ as long as $P(x) \neq \emptyset$. Then, there is a maximal element $x^* \in X$, $P(x^*) = \emptyset$.*

Proof. Assume the contrary. Then, for all $x \in X$, there is a neighbourhood U_x of x and an upper semicontinuous correspondence $\theta_x : U_x \to E'_X$ satisfying that $\langle p, w - z \rangle > 0$ for all $p \in \theta(z)$ and $w \in P(z)$. Since X is compact, there is a finite covering $U_{x^1}, \ldots, U_{x^m}$ of X, so that by using the partition of unity $\alpha^t : U_{x^t} \to [0, 1]$, $t = 1, \ldots, m$ subordinated to $U_{x^1}, \ldots, U_{x^m}$, we obtain an upper semicontinuous correspondence $\theta : X \to E'_X$ as

$$\theta(x) = \sum_{t=1}^{m} \alpha^t(x) \theta_{x^t}(x),$$

where $\theta_{x^t}(x)$ is defined to be $\{0\}$ for all $x \notin U_{x^t}$. Since $\theta : X \to E'_X$ is upper semicontinuous and compact valued, and since the bilinear form $\langle \cdot, \cdot \rangle$ on $X \times E'_X$ is jointly continuous, the direction structure $V(x, y)$ on X defined by $V(x, y) = \{w \in X | \langle p, w - x \rangle > 0\}$ if $y \in \{w \in X | \langle p, w - x \rangle > 0\}$, else $V(x, y) = \emptyset$, is lower topological and P has a locally fixed direction at each $x \in X$ under V. Hence, by Theorem 3, P has a fixed point. Since $\langle p, w - z \rangle > 0$ for all $p \in \theta(z)$, $w \in P(z)$, and $z \in U_x$ as long as $P(x) \neq \emptyset$, P cannot have a fixed point, so that we have a contradiction. □

We next consider the problem on the existence of maximal elements among many agents, i.e., the social optima. Adding to (B0), we use the following settings:

[17] The assumed property for the bilinear form may be satisfied under the standard situation for the commodity-price duality in economic equilibrium theory. For example, given a duality (E, F), suppose that E has the Mackey topology $\tau(E, F)$ and E'_X is a dense subset of a $\sigma(F, E)$ compact subset D of F. The joint continuity for the bilinear form has an economic meaning that we are considering the situation in which with simultaneous small changes in prices and amounts for commodities associate small changes in their total values.

(B1) [Consumers and Producers]: Let $X_1,\ldots,X_m$ and $Y_1,\ldots,Y_n$ be compact convex subsets of X. Moreover, let $\omega^1,\ldots,\omega^m$ are points in E such that $\sum_{i=1}^m X_i \cap (\sum_{j=1}^n Y_j + \sum_{i=1}^m \omega^i) \neq \emptyset$.

We call the set $\mathcal{Alloc} = \prod_{i=1}^m X_i$ the *set of allocations* (for consumers) and the set $\mathcal{Falloc} = \{((x^i)_{i=1}^m) \in \mathcal{Alloc} | \sum_{i=1}^m x^i = \sum_{j=1}^n y^j + \sum_{i=1}^m \omega^i, (y^j)_{j=1}^n \in \prod_{j=1}^n Y_j\}$ the *set of feasible allocations*. Under (B1), we see that $\mathcal{Falloc}$ is a non-empty compact convex subset of $\mathcal{Alloc}$.

Together with (B0) and (B1), let us assume that for each $i = 1,\ldots,m$, there are two correspondences $P_i : \mathcal{Alloc} \to X_i$ and $\tilde{P}_i : \mathcal{Alloc} \to X_i$, the *strong* and *weak preference correspondences*, respectively, satisfying that $\forall x = (x^1,\ldots,x^m) \in \mathcal{Alloc}$, $x^i \notin P_i(x)$ (the irreflexivity), $x^i \in \tilde{P}_i(x)$ (the reflexivity), and $P_i(x) \subset \tilde{P}_i(x)$. Then, let us define a correspondence, $P : \mathcal{Alloc} \to \mathcal{Alloc}$, as follows:

$$P(x) = \{w = (w^1,\ldots,w^m) | \forall i, w^i \in \tilde{P}_i(x) \text{ and } \exists i, w^i \in P_i(x)\}$$

For allocations $x, y \in \mathcal{Alloc}$, y is said to be *Pareto superior* to x if and only if the condition $y \in P(x)$ is satisfied. It is easy to check that for all $z \in \mathcal{Alloc}$, $z \notin P(z)$, i.e., $P : \mathcal{Alloc} \to \mathcal{Alloc}$ is an irreflexive correspondence. A feasible allocation $x \in \mathcal{Falloc}$ is said to be *Pareto optimal* if $P(x) \cap \mathcal{Falloc} = \emptyset$. The next theorem shows that the existence of Pareto optimal allocations may be assured even for cases with non-ordered, non-convex, and non-continuous individual preferences. Note that the necessary condition for individual preferences to assure the existence of social optima (Pareto optimal allocations), (D), is essentially the same as the necessary condition for the existence of individual optima (maximal elements) in Theorem 10.

Theorem 5. *(Existence of Pareto Optimal Allocations) Assume that for each $i = 1,\ldots,m$, P_i and $\tilde{P}_i$ satisfies the following conditions for directions of mappings.*

> *(D) For each $x \in \mathcal{Alloc}$, (resp., for each $x \in \mathcal{Alloc}$ such that $P_i(x) \neq \emptyset$), there are a neighbourhood U_x of x and an upper semicontinuous compact valued correspondence $\theta_x^i : U_x \to E'_X$ such that $\langle p, w_i - z_i\rangle \geq 0$ (resp., $\langle p, w_i - z_i\rangle > 0$) for all $p \in \theta_x^i(z)$, $w_i \in \tilde{P}_i(z)$, (resp., $w_i \in P_i(z)$), and $z = (z_1,\ldots,z_m) \in U_x$.*

Then, there is a Pareto optimal allocation.

Proof. Assume the contrary. Then, the correspondence $P_F : \mathcal{Falloc} \ni x \mapsto P(x) \cap \mathcal{Falloc}$ is non-empty valued. By condition (D) and the compactness of $\mathcal{Alloc}$, through the argument using the partition of unity, we obtain an upper semicontinuous compact valued correspondence $\theta^i : \mathcal{Alloc} \to E'_X$ for each i such that $\forall z \in \mathcal{Alloc}$ and $\forall p \in \theta^i(z)$, $(w_i \in \tilde{P}_i(z)) \to (\langle p, w_i - z_i\rangle \geq 0)$

and $(w_i \in P_i(z)) \rightarrow (\langle p, w_i - z_i \rangle \geq 0))$. For each i and $z \in \boldsymbol{Alloc}$, denote by $\hat{P}_i(z)$, (resp., $\hat{\hat{P}}_i(z)$), the set $\{w_i \in X_i | \forall p \in \theta^i(z), \langle p, w_i - z_i \rangle > 0\}$, (resp., the set $\{w_i \in X_1 | \forall p \in \theta^i(z), \langle p, w_i - z_i \rangle \geq 0\}$). Clearly, for each i and $z \in \boldsymbol{Alloc}$, $\hat{P}_i(z) \supset P_i(z)$ and $\hat{\hat{P}}_i(z) \supset \tilde{P}_i(z)$. Since $\hat{P}_i(z)$ is open, and since the sum operation is continuous, the non-emptiness of P_F means that for each $w \in \mathcal{F}\boldsymbol{alloc}$, there is an neighbourhood O_w in $\mathcal{F}\boldsymbol{alloc}$ and an index of consumer $i(w)$ such that for each $z \in O_w$, there is an element $y = (y_1, \ldots, y_m) \in \mathcal{F}\boldsymbol{alloc}$ such that $y_{i(w)} \in \hat{P}_{i(w)}(z)$. Since $\mathcal{F}\boldsymbol{alloc}$ is compact, there is a finite covering $O_{w^1}, \ldots, O_{w^k}$ of $\mathcal{F}\boldsymbol{alloc}$ and the partition of unity, $\alpha_t : O_{w^t} \rightarrow [0,1]$, $t = 1, \ldots, k$, subordinated to it. Let us define a correspondence on $\mathcal{F}\boldsymbol{alloc}$ to $\mathcal{F}\boldsymbol{alloc}$, Φ as, for each $z = (z_1, \ldots, z_m) \in \mathcal{F}\boldsymbol{alloc}$,

$$\Phi(z) = z + \alpha^t(z)(\Phi(z)^t - z), \text{ where}$$

$$\Phi^t(z) = \hat{\hat{P}}_1(z) \times \cdots \times \hat{\hat{P}}_{i(w^t)-1}(z) \times \hat{P}_{i(w^t)}(z) \times \hat{\hat{P}}_{i(w^t)+1} \times \cdots \times \hat{\hat{P}}_m(z) \cap \mathcal{F}\boldsymbol{alloc}.$$

Φ is convex valued since each Φ^t is. Φ has non-empty valued since each Φ^t is non-empty valued as long as P_F is. Moreover, by defining $\theta(z) : \boldsymbol{Alloc} \rightarrow \boldsymbol{Alloc}$ and $\Psi : \mathcal{F}\boldsymbol{alloc} \rightarrow \mathcal{F}\boldsymbol{alloc}$, respectively, as

$$\begin{aligned} \theta(z) &= (\theta^1(z), \ldots, \theta^m(z)), \text{ and} \\ \Psi(z) &= \{w \in \mathcal{F}\boldsymbol{alloc} | \forall p \in \theta(z), \langle p, w \rangle > 0\}, \end{aligned}$$

we have $\Phi(z) \subset \Psi(z)$, so that the correspondence, Ψ, is a non-empty convex valued on $\mathcal{F}\boldsymbol{alloc}$ to $\mathcal{F}\boldsymbol{alloc}$ having no fixed point. Note that θ is also compact valued upper semicontinuous correspondence on $\boldsymbol{Alloc}$ to ${E'_X}^{(m)}$, where ${E'_X}^{(m)}$ denotes the m-th product of E'_X. Then, for all $w^* \in \Psi(z^*)$, there is an $\epsilon^* > 0$ such that $\langle p, w^* \rangle > \epsilon^* > 0$ by the compactness of $\theta(z^*)$, hence, by the joint continuity of the duality operation, (precisely, for each component,) there is a neighbourhood O^* of z^* such that for all $z \in O^*$, $w^* \in \Psi(z)$. Therefore, Ψ has a locally continuous (fixed) selection at each z, so that by Corollary 6.9, Ψ has a fixed point and we have a contradiction. □

The set E'_X may be interpreted as the set of *prices*. For each price $p \in E'_X$, we may define a non-empty set of *excess demands*, $\zeta(p) \subset X$, as the consequence of agents' maximization behaviors[18]. If we add the following conditions for the excess demand correspondence, ζ, we may argue for the market equilibrium and the existence of equilibrium prices.

---

[18] Under (B0) and (B1), if we suppose the ordinary private ownership structure as in Debreu (1959), Theorem 10 clearly assures the non-emptiness for each agent's optimal behaviors, $\zeta(p)$. Conditions (B2) and (B3) below may also be satisfied. For condition (E) in the next theorem, however, we do not have such an individual preference foundation. In this sense, it is more natural for us to consider the excess demand correspondence, $p \mapsto \zeta(p)$, as a primitive notion (the excess demand ap-

(B2) [Walras' Law]: For each $p \in E'_X$, $\langle p, z\rangle = 0$ for all $z \in \zeta(p)$.
(B3) If $0 \notin \zeta(p)$, there exists $q \in E'_X$ such that $\langle q, z\rangle > 0$ for all $z \in \zeta(p)$.

Condition (B2) is always satisfied as long as every consumer satisfies their ordinary budget constraint with equality. Condition (B3) says that the price set is taken to be sufficiently large so that we may adjust any non-zero amount of excess demands. We say that the price p is *adjusted* by price q if the relation in (B3), $\langle q, z\rangle > 0$ for all $z \in \zeta(p)$, is satisfied. By (B2), p is not adjusted by p for all $p \in E'_X$. A price, $p^* \in E'_X$ is said to be an *equilibrium price* if $0 \in \zeta(p^*)$. Under (B3), a price p^* is an equilibrium price if and only if p^* is not adjusted by q for all $q \in E'_X$. The mathematical result on the existence of equilibrium for this type of abstract settings is known as Gale-Nikaido-Debreu theorem [19]. In our general settings, the theorem may be extended as follows.

Theorem 6. *(Existence of Price Equilibrium) Assume (B0), and let us consider an excess demand correspondence, $\zeta : E'_X \to X$, satisfying (B2) and (B3). Moreover, suppose that on the closure D of E'_X, condition (T) in Corollary 6.7 is satisfied, and that ζ has a continuous local direction at each non-equilibrium point in the following sense.*

(E) For each r in D, if r is not an equilibrium price, there is a neighbourhood U_r of r in D and a continuous function $q_r : U_r \to E'_X$ such that for all $p \in U_r \cap E'_X$, $\langle q_r, z\rangle > 0$ for all $z \in \zeta(p)$.

Then, there is an equilibrium price $p^* \in E'_X$, $0 \in \zeta(p^*)$.

Proof. Assume the contrary, so that there is no $p \in E'_X$ such that $0 \in \zeta(p)$. (Note that the closure, D, of E'_X in E^* is compact and convex.) Then for all r in D, there is a neighbourhood U_r of r in D and a continuous function $q_r : U_r \to E'_X$ satisfying the condition stated above. Since the closure of E'_X is compact, there are finite subcovering $U_{r^1}, \ldots, U_{r^k}$ of D. Let $\alpha^t : U_{r^t} \to X$, $t = 1, \ldots, k$ be the partition of unity subordinated to U_{r^t}, $t = 1, \ldots, k$. Then, the mapping $f : D \ni r \mapsto \sum_{t=1}^{k} \alpha^t(r) q_{r^t}(r) \in D$ is continuous and has no fixed point since for all t such that $p \in U_{r^t}$, $\langle q_{r^t}, z\rangle > 0$ for all $z \in \zeta(p)$ though $\langle p, z\rangle$ should be 0 under (B2). Since condition (T) on D is satisfied, f has a fixed point by Corollary 6.7, so that we have a contradiction. □

In an extension of Gale-Nikaido-Debreu theorem, condition (E) may be considered as one of the weakest requirements for the continuity of excess de-

proach). Needless to say, condition (E) is more general than to assume the ordinary situation such that the excess demand correspondence is compact convex valued and upper semi-continuous.

[19] See, for example, Debreu (1956). For one of the most general treatments in locally convex space with acyclic valued correspondences, see Nikaido (1957) and Nikaido (1959), Section 5.2.

mand correspondences. Except for directions of mappings, there is no topological requirement for the set of values at each point [20]. It should also be noted, however, that condition (E) includes the so called *boundary condition* for the excess demand correspondence (on the boundary $D \setminus E'_X$ of E'_X in E^*). As a boundary condition, (i.e., as a condition for points $r \in D \setminus E'_X$,) condition (E) may not be called the weakest one [21]. There is an alternative way for assuring the existence of equilibrium prices under more general conditions for boundary points, though requirements for the continuity in (E) should be strengthened.

If we write $B(p) = \{q \in E'_X | \exists z \in \zeta(q), \langle p, z \rangle \leq 0\}$ for each $p \in E'_X$, (the set of price q which may not be adjusted by price p), then the following (E') (a slight modification of the continuity condition in (E)) is a sufficient condition for $B(p)$ to be a closed subset of E'_X.

> (E') For each q in E'_X, if q is not an equilibrium price, there is a neighbourhood U_q of q in E'_X and a point $p_q \in E'_X$ such that for all $q' \in U_q$, $\langle p_q, z \rangle > 0$ for all $z \in \zeta(q')$.

It is also easy to check that for each finite set $p^1, \ldots, p^k$, $\mathrm{co}\,\{p^1, \ldots, p^k\} \subset \bigcup_{t=1}^k B(p^t)$. Hence, by using Knaster-Kuratowski-Mazurkiewicz theorem together with the following boundary condition, we also obtain the existence of equilibrium prices.

> (B4) [Boundary Condition]: For all r in the boundary, $\partial E'_X$, of $E'_X \subset D$ and for all net $\{p^\nu, \nu \in \mathfrak{N}\}$ in E'_X converging to r, there are an element $q_r \in E'_X$ and a subnet $\{p^{\nu(\mu)}, \mu \in \mathfrak{M}\}$ of $\{p^\nu, \nu \in \mathfrak{N}\}$ such that $\langle q_r, z \rangle > 0$ for all $z \in \zeta(p^{\nu(\mu)})$ for all $\mu \in \mathfrak{M}$.

In this case, the generality of the continuity condition (E') together with the boundary condition (B4) is essentially the same with the condition used in the market equilibrium existence theorem in Urai (2000), Theorem 8.

It is also possible to apply theorems in the previous section and settings (B1), (B2) to the existence of Nash equilibrium and Generalized Nash equilibrium for an abstract economy of the Arrow-Debreu type. In such cases, we obtain various results on the existence of equilibrium with (possibly) non-ordered, non-convex, and/or non-continuous preferences and constraint correspondences in Hausdorff topological vector spaces. For some results, see Urai and Yoshimachi (2002).

---

[20] Of course, it is easy to check that every non-empty closed convex valued upper semicontinuous correspondence satisfies condition (E).

[21] In condition (E), the boundary behavior is described for a neighbourhood of each boundary point. Boundary conditions of the weakest type usually treat each boundary point independently. See, for example, Aliprantis and Brown (1983), Mehta and Tarafdar (1987), etc. For a more general and unified treatment, see Urai (2000), Section 4.

References

Aliprantis, C.D., Brown, D.J.: Equilibria in markets with a Riesz space of commodities. Journal of Mathematical Economics **11**, 189-207 (1983)

Arrow, K.J., Debreu, G.: Existence of an equilibrium for a competitive economy. Econometrica **22**, 265-290 (1954)

Browder, F.: The fixed point theory of multi-valued mappings in topological vector spaces. Mathematical Annals **177**, 283-301 (1968)

Debreu, G.: Market equilibrium. Proceedings of the National Academy of Sciences of the U.S.A. **42**, 876-878 (1956), Reprinted as Chapter 7 in G. Debreu, Mathematical Economics, Cambridge University Press, Cambridge 1983

Debreu, G.: Theory of Value. Yale University Press, New Haven, CT 1959

Gale, D., Mas-Colell, A.: An equilibrium existence theorem for a general model without ordered preferences. Journal of Mathematical Economics **2**, 9-15 (1975), (For some corrections see Journal of Mathematical Economics **6**, 297-298 (1979))

Mehta, G., Tarafdar, E.: Infinite-dimensional Gale-Nikaido-Debreu theorem and a fixed-point theorem of Tarafdar. Journal of Economic Theory **41**, 333-339 (1987)

Nikaido, H.: On the classical multilateral exchange problem. Metroeconomica **8**, 135-145 (1956), A supplementary note: Metroeconomica **8**, 209-210 (1957)

Nikaido, H.: Existence of equilibrium based on the Walras' law. ISER Discussion Paper No. 2, Institution of Social and Economic Research 1957

Nikaido, H.: Coincidence and some systems of inequalities. Journal of The Mathematical Society of Japan **11**(4), 354-373 (1959)

Schaefer, H.H.: Topological Vector Spaces. Springer-Verlag, New York/Berlin 1971

Shafer, W., Sonnenschein, H.F.: Equilibrium in abstract economies without ordered preferences. Journal of Mathematical Economics **2**, 345-348 (1975)

Urai, K.: Fixed point theorems and the existence of economic equilibria based on conditions for local directions of mappings. Advances in Mathematical Economics **2**, 87-118 (2000)

Urai, K., Hayashi, T.: A generalization of continuity and convexity conditions for correspondences in the economic equilibrium theory. The Japanese Economic Review **51**(4), 583-595 (2000)

Urai, K., Yoshimachi, A.: Existence of equilibrium with non-convex constraint correspondences: Including an application for the default economy. Discussion Paper No. 02-06, Faculty of Economics and Osaka School of International Public Policy, Osaka University 2002

Adv. Math. Econ. 6, 167–183 (2004)

Advances in
MATHEMATICAL
ECONOMICS

Monetary equilibrium with buying and selling price spread without transactions costs*

Akira Yamazaki†

Graduate School of Economics, Hitotsubashi University, Kunitachi, Tokyo, 186-8601, Japan

Received: July 22, 2003
Revised: October 8, 2003

JEL classification: D51, E40

Mathematics Subject Classification (2000): 91 B

Abstract. This paper is concerned with the Hahn problem in a general monetary equilibrium model at the terminal period. Under the assumption that an initial endowment allocation is not Pareto optimal it is proved that an equilibrium with a positive value of money exists if traders take buying and selling prices of commodities as given even if transactions costs are not explicitly required in the buying and selling activities of traders in commodity markets.

This result seems to suggest two interpretations. One is that a standard monetary equilibrium concept must be strengthened so as to explicitly require an arbitrage-free property of bid-ask spreads of commodity prices vis-à-vis transactions costs. The second interpretation is that a model in which traders take distinct buying and selling prices as given although no transactions costs are required can be thought of as a way to make tax payments required by an external authority in classical papers endogenous in the form of an indirect taxation.

Key words: general equilibrium model, monetary equilibrium, Hahn problem, buying and selling price spread; transactions costs

---

* An earlier version of this paper was circulated as "On the Terminal Value of Money without Transactions Costs", RUEE Working Paper #91-44, Department of Economics, Hitotsubashi University, October, 1991.

† I would like to thank an anonymous referee for his/her useful comments on an earlier draft of this paper.

1. Introduction

This paper is concerned with the Hahn problem in a general monetary equilibrium model at the terminal period[1] (see Hahn (1965) and Duffie (1990)). Under the assumption that an initial endowments allocation is not Pareto optimal it is proved that an equilibrium with a positive value of money exists if traders take buying and selling prices of commodities as given even if transactions costs are not explicitly required in the buying and selling activities of transactions in commodity markets.

When one introduces pure outside fiat money into a standard general equilibrium model of Arrow-Debreu-McKenzie-Nikaido (ADMN), the fact that an equilibrium exists in the original ADMN model implies that there always exists an equilibrium in which the value of money is zero. But, of course, unless money receives positive valuation in markets, it is impossible to play a basic role as a means of payment. This is the well-known "Hahn problem" in a general monetary equilibrium model pointed out by Hahn (1965). In a series of efforts to solve the problem in 1970's, traders are either simply forced to hold the positive amount of money or motivated to such holdings by required tax payments in the terminal period. (See, e.g., Starr (1974), Kurz (1974a, 1974b), Heller (1974), Okuno (1973), etc.) The reason for such a requirement is that fiat money has no apparent purchasing power in the last period so that traders have no incentives to hold money.

Duffie (1990) in his contribution to this problem has specifically taken up the problem of terminal value of money. In the setting of an ADMN model with outside fiat money and transactions costs expressed by individual transactions possibility sets as in Kurz (1974b) and Heller (1974), he showed that an equilibrium with the positive terminal value of money exists if traders take buying and selling prices in markets as given provided that they have incentives to trade (that is, the initial endowments allocation is not Pareto optimal). Traders are thought to face distinct buying and selling prices due to transactions costs as in a market transactions costs' model of Foley (1970). Duffie's basic idea is that accounting identity forces the total value of purchases to exceed the total value of sales by the amount of outside money if traders are to use the money balances in their transactions at the terminal period.

An equilibrium is composed of buying prices, selling prices, and preference-maximizing transactions in traders' budgets given these prices such that the total amount of each commodity (or money) bought does not exceed the total amount of each commodity (or money) sold. This equilibrium concept was introduced by Foley (1970) and Hahn (1971), and others such as Kurz

[1] The "Hahn problem" addresses the issue of whether there exists an equilibrium in which the value of money is positive.

(1974a, 1974b), Starrett (1973), Hayashi (1974), Okuno (1974), Duffie (1990), etc. worked with this definition. In the present paper we employ an exchange economy version of a standard ADMN model as in Duffie (1990) but without explicitly introducing transactions costs in the form of individual transactions possibility sets as is done in his paper as well as in the works of Kurz (1974b) and Heller (1974). The model describes the terminal period of a finite sequence economy. We work with the above equilibrium concept. An existence theorem presented confirms a conjecture of Duffie (1990) that his existence theorem does not seem to rely on the requirement of transactions costs.

Did something go wrong in the setting of transactions costs models? Or, how should we interpret the implication of the result obtained in this paper? There seem to be two possible interpretations. One interpretation is that the equilibrium concept needs to be strengthened so that bid-ask spreads of buying and selling prices do satisfy an arbitrage-free condition defined in an appropriate way vis-à-vis explicitly introduced transactions costs. The second interpretation is to regard bid-ask spreads as representing endogenous indirect taxation. The latter interpretation may be of interest as writers such as Lerner (1947), Starr (1974), Kurz (1974b) and others thought that it is necessary to require that terminal money be used for tax purposes to force traders to demand the terminal money. The formulation of taxes by distinct buying and selling prices without transactions costs do endogenize this taxation.

Finally, let us note that the definition of an equilibrium given above presupposes the free disposability of commodities. By allowing negative buying and selling prices, we do not assume the free disposability. The sign of a bid-ask spread then depends on the sign of a buying price (see Section 3 on this). We exploit the proof technique of Bergstrom (1976) and Shafer-Sonnenschein (1975) together with the idea of Duffie (1990).

2. The model

Consumption characteristics of agents

The commodity space is $\mathbb{R}^\ell$. For simplicity consumption sets are taken to be $\mathbb{R}^\ell_+$. Preference relations $\succ \subset \mathbb{R}^\ell_+ \times \mathbb{R}^\ell_+$ are relatively open, irreflexive, locally nonsatiated, and convex (i.e., the set $\{z \in \mathbb{R}^\ell_+ \mid z \succ x\}$ is convex for each $x \in \mathbb{R}^\ell_+$).

There are m agents. Each agent $i \in \{1, \ldots, m\}$ is characterized by a preference relation $\succ_i$, initial endowments $e_i \in \mathbb{R}^\ell_+$, and an endowment of money $M_i \geq 0$.

Market transactions

Buying and selling are separate transaction activities which command different price systems. All the prices are expressed in terms of the unit of account. $p^B \in \mathbb{R}^\ell$ and $p^S \in \mathbb{R}^\ell$ denote buying and selling prices respectively. Some of the prices may be negative. We express a pair composed of buying prices p^B and "reverse-signed" selling prices $-p^S$ by $p := (p^B, -p^S)$. p represents a vector of prices to be *paid* for a unit of purchases or sales of commodities in the market. A basic assumption is that each agent takes both buying and selling prices in markets as given and determines its transaction of commodities.

Means of payment

Means of payment are fiat money the amount $M = \sum_{i=1}^m M_i > 0$ of which is fixed exogenously. It is implicitly assumed that a public authority is set up for coordinating market transactions and collecting money for its service provided in the markets. The monetary unit is used as the unit of account.

Individual transaction

Let x_i^B, x_i^S $(\in \mathbb{R}_+^\ell)$ represent purchases and sales of ℓ commodities by agent i in the markets. Write $x_i := (x_i^B, x_i^S) \in \mathbb{R}_+^\ell \times \mathbb{R}_+^\ell$. The transaction of x_i by agent i leaves the consumption vector $x_i^C := x_i^B - x_i^S + e_i$ for i. It is *budget feasible* for i at buying and selling prices $p = (p^B, -p^S) \in \mathbb{R}^{2\ell}$ if $x_i^C \geq 0$ and $p \cdot x_i \leq M_i$. When agent i engages in market transactions x_i, he receives $p^S \cdot x_i^S$ in money and spends at most $p^S \cdot x_i^S + M_i$. Or, one can think of the above budget inequality as allowing netting between obligations and claims before the payment of money is made to the market authority.

A budget feasible transaction x_i is a *preference maximizer* for i provided $y_i^C \not\succ_i x_i^C$ for any budget feasible transaction y_i for i.

Allocations and monetary equilibrium

A m-tuple of transaction vectors $(x_1, \ldots, x_m)$ is a *transaction allocation* if $x_i \in \mathbb{R}_+^{2\ell}$ and $x_i^C \in \mathbb{R}_+^\ell$ for every $i = 1, \ldots, m$. $(x_1^C, \ldots, x_m^C)$ is called the associated *consumption allocation.* A transaction allocation is *feasible* if $\sum_{i=1}^m x_i^B = \sum_{i=1}^m x_i^S$. Note that the feasibility of a transaction allocation is equivalent to the feasibility condition $\sum_{i=1}^m x_i^C = \sum_{i=1}^m e_i$ of the associated consumption allocation $(x_1^C, \ldots, x_m^C)$, which is more familiar.

Given two transaction allocations $(x_1, \ldots, x_m)$ and $(y_1, \ldots, y_m)$, $(x_1, \ldots, x_m)$ is said to *Pareto improve* $(y_1, \ldots, y_m)$ if $x_i^C \succ_i y_i^C$ for every agent i. (This requirement of improvement is stronger than the usual one unless preferences satisfy monotonicity.) A feasible transaction allocation $(x_1, \ldots, x_m)$ is *Pareto optimal* if no other feasible transaction allocations can Pareto improve it.

A *monetary equilibrium* for the economy $(\succ_i, e_i, M_i)_{i=1,\ldots,m}$ is a collection $((x_1, \ldots, x_m), p) \in (\mathbb{R}^{2\ell})^m \times \mathbb{R}^{2\ell}$ such that, given prices p^B, p^S, the transaction x_i is budget feasible and a preference maximizer among budget feasible transactions for each $i \in \{1, \ldots, m\}$, and the transaction allocation $(x_1, \ldots, x_m)$ is feasible, i.e., $\sum_{i=1}^m x_i^B = \sum_{i=1}^m x_i^S$.

In this definition of a monetary equilibrium buying and selling prices are expressed in terms of the unit of account. Thus, whenever equilibrium prices $p \in \mathbb{R}^{2\ell}$ exist, the value of money must be positive. One may note that in seeking candidate equilibrium prices of commodities they cannot be constrained to lie in a compact subset of $\mathbb{R}^{2\ell}$ unless the value of money which might fall to zero is explicitly introduced. It is to be noted that the free disposability of commodities is not assumed so that some of the prices may be negative at equilibrium.

Let us give a statement of a property concerning the preference distribution of an economy which will be needed as a condition of the theorem below.

> [*Possibility of Individual Utility Enhancement for Feasible Allocations*] There is a positive number k^* such that for any given feasible consumption allocation $(x_1^C, \ldots, x_m^C)$, $\sum_{i=1}^m x_i^C = \sum_{i=1}^m e_i$, and for any commodity j, $(1 \leq j \leq \ell)$, there is some agent $i \in \{1, \ldots, m\}$ with a consumption vector $0 \leq y \leq k^* \sum_{i=1}^m e_i$ satisfying
> (i) $y^h = x_i^{Ch}$ for all $h \neq j$,
> (ii) $y \succ_i x_i^C$.

Take any one commodity j. If it is distributed freely among agents exhausting all of the resource, then, it should be the case that some one prefers to reduce or increase (within some uniform bound) his/her consumption of the commodity. If there is at least one agent who consumes and regards the commodity j to be desirable (i.e., a "good") or undesirable (i.e., a "bad") in all the range of feasible consumptions, this condition is automatically satisfied. Thus, it is immediate that if all the agents have monotone preferences, the above property is satisfied. Note that it does not rule out some of the commodities to be bads or some of the commodities to become undesirable beyond some levels of their consumption as long as all the agents do not reach their satiation level of a particular commodity simultaneously.

The only case of preference distribution which is ruled out by this condition is the following: There is a commodity j that is regarded desirable only up to some positive consumption levels by *all* agents, and the sum of the "satiation levels" of the commodity, beyond which an increase of its consumption by agents become undesirable just happens to be exactly equal to the total endowment of the commodity j in the economy. This condition is of a "generic" nature in the sense that even if a particular economy does not satisfy the condition, a slight perturbation of agents' preferences or total endowments will make the condition satisfied.

Theorem *Let* $(\succ_i, e_i, M_i)_{i=1,\ldots,m}$ *be an economy with* $\sum_{i=1}^{m} e_i > 0$ *and* $M = \sum_{i=1}^{m} M_i > 0$ *satisfying the property of the possibility of individual utility enhancement for feasible allocations. Then, there is a monetary equilibrium* $((x_1, \ldots, x_m), p)$ *for the economy provided that the transaction allocation* $(0, \ldots, 0)$ *inducing the initial endowments consumption allocation* $(e_1, \ldots, e_m)$ *is not Pareto optimal.*

3. Discussion and some remarks

A basic scenario supporting the formulation of the model described in the previous section is essentially the same as the one given in Duffie (1990) except for the explicit statement of transactions costs in terms of individual transaction possibility sets in case of Duffie.

A market authority is organized and performs the operation of exchanging ℓ commodities for money at each point in time for a finite duration. The model describes the terminal period only. The market authority plays the role of a "central banker" and collects money at the time of transactions. Fiat money need not be paper currency but could take the form of "electro-money" in the sense that it is composed of accounts held by traders at the central bank with debits and credits done by electronic devices. Overdrafts are not permitted and we require non-negative balances of money by traders.

Aside from the fact that money being the unit of account, a basic role of money in the model is transactional one at the terminal period. Although it is implicit, money also plays the role of a store of value carried over to the terminal period from the implicit previous time periods in the form of outside money balance.

The theorem asserts that if traders take buying and selling prices of commodities in markets as given, then there is a competitive market equilibrium with a positive monetary value even if no transaction costs are required provided that traders are better off with trades than without trades. This result confirms the conjecture of Duffie that his theorem establishing the existence of monetary equilibrium may obtain without transactions costs (see Duffie (1990, Theorem 1, p.87, and the second paragraph in p.92)).

Let us briefly indicate the nature of buying and selling prices of commodities and the value of money at equilibrium that are given in the proof of the theorem. In our search for equilibrium buying and selling prices and the value of money we normalize buying prices p^B to the unit ball $\mathbb{B}^\ell$ in $\mathbb{R}^\ell$ and express selling prices p^S and value of money p^M relative to p^B. The concept of volume of trades is introduced as in Duffie (1990) to determine selling prices. The volume of trades v associated with a transaction allocation $(x_1, \ldots, x_m)$ is defined as the vector composed of the maximum number of total purchases or sales of each commodity at markets, that is,

$$v := \max\left\{\sum_{i=1}^m x_i^B, \sum_{i=1}^m x_i^S\right\} \in \mathbb{R}_+^\ell$$

with the maximum taken coordinatewise. At equilibrium we have $\sum_{i=1}^m x_i^B = \sum_{i=1}^m x_i^S = v$.

Given volume of trade $v \in \mathbb{R}_+^\ell$, selling prices p^S associated with buying prices p^B are defined via bid-ask spread factor $\delta^j(p^B, v)$ given by

$$\delta^j(p^B, v) := \frac{sgn(p^{Bj})v^j}{1 + v^j}$$

for $j = 1, \ldots, \ell$. The bid-ask spread of commodity j is then $\delta^j(p^B, v)p^{Bj}$. In other words, the selling price p^{Sj} of commodity j is given by $(1-\delta^j(p^B, v))p^{Bj}$. The absolute value of bid-ask spread factor of a commodity is strictly less than one, and it is monotonic in volumes of trades of that commodity. One might feel that the latter property is somewhat counter to what one might expect as a property of bid-ask spreads. Note, however, that the spirit of bid-ask spreads in the present model is that they originate in transactions costs which certainly accumulate as volumes of trades increase. Bid-ask spreads are designed so that a positive buying price commands a positive spread and a negative price a negative spread. This means that for desired commodities buyers pay more than sellers receive, and for undesired commodities, in order to have them disposed, sellers pay more than buyers receive.

During the course of the existence proof in Section 4, given buying prices p^B, selling prices p^S are determined according to bid-ask spreads given in the previous paragraph. The value p^M of money is set so that the value of existing stock of outside money is exactly equal to the total sum of the bid-ask spreads resulting from the volume v of trades in markets. It means that agents use outside money to finance the excess of their payments over receipts in order to clear the results of their transactions with the market authority in commodity markets.

By the hypothesis of the theorem the initial endowments allocation is not Pareto optimal. Thus, agents do wish to engage in trades provided bid-ask spreads are sufficiently small. Bid-ask spreads introduced above have the property that they can be made arbitrarily small in neighborhoods of the initial endowments allocation. It therefore follows that there will be a positive volume of trades at equilibria. By the way value of money is determined in our model, a positive volume of trades induces a positive terminal value of money.

Nevertheless, the theorem may at first seem counterintuitive. Since the model deals with the terminal period, it must be transaction demands that give rise to the positivity of monetary value. Market transactions of commodities must require costs if they are to motivate transactions demands for money. Thus, a natural question might arise: Why is it that one can establish the existence of monetary equilibrium without requiring transactions costs? The answer is straightforward. It is because we assumed that agents take buying and selling prices as given no matter how big bid-ask spreads are. But agents will in general take advantage of bid-ask spreads unless their arbitrage transactions are unprofitable due to existing transactions costs. Therefore, the fact that the theorem is true seems to point to the need for strengthening the equilibrium concept adopted by us and others in the context of a general equilibrium model with transactions costs. It is implicitly assumed in the literature (see, e.g., Foley (1970) and Duffie (1990, p.90)) that transactions costs are severe enough to prevent arbitrage over bid-ask spreads. What need to be done thus seems to be to require explicitly bid-ask spreads of prices be arbitrage free in the sense that arbitrage transactions are not profitable vis-à-vis transactions costs incurred by such activity. It would lead to rethinking of the formulation of the model with individual transactions technologies.

We would like to note that writers such as Lerner (1947), Hahn (1971), Starr (1974), Heller (1974), and others have either simply forced agents to demand fiat money at the terminal period or motivated such holdings requiring the payment of exogenously given lump sum taxes. In this context we could regard bid-ask spreads of prices in the present model as *endogenous indirect taxes* by the authority. Spread factor δ^j then represents a quantity tax levied on commodity j. One is thus led to feel that it is natural to have existence of a monetary equilibrium in a framework without explicit transactions costs. And a conceptual difficulty pointed out in the previous paragraph would not arise in this interpretation.

A final remark may be due. The model in this paper does not suggest why fiat money may offer transactional advantages as a medium of exchange, but given a positive value for money the advantages are not difficult to imagine, considering for instance the work of Kiyotaki and Wright (1989), and Ostroy and Starr (1973). Instead the model presented here merely suggests why outside fiat money may have positive value in the first place even in a situation where explicit transactions costs do not exist.

4. Proof of the Theorem

1. Let k^* be the positive number given in the statement of possibility of utility enhancement for feasible allocations, and let

$$\bar{e} := \max\{k^*, 2\}\sum_{i=1}^{m} e_i > 0\,.$$

For each $i = 1, \dots, m$, define

$$X_i := [0, \bar{e}] \times [0, \bar{e}] \subset \mathbb{R}^\ell_+ \times \mathbb{R}^\ell_+$$

where $[0, \bar{e}] := \{z \in \mathbb{R}^\ell |\, 0 \le z^j \le \bar{e}^j \text{ for all } j = 1, \dots, \ell\}$. Let us also define

$$X := \prod_{i=1}^{m} X_i \text{ and } V := [0, \bar{e}] \subset \mathbb{R}^\ell_+ .$$

X is the space of transaction allocations and an element $v \in V$ will be called *volume of trades* which will be defined later.

$\mathbb{B}^\ell$ denotes the unit closed ball in $\mathbb{R}^\ell$ centered at the origin 0, and a vector q in $\mathbb{B}^\ell$ will represent buying prices of ℓ commodities. Given buying prices q and volume of trades v, selling prices are defined via a *spread factor* $\delta^j(q, v)$, $j = 1, \dots, \ell$, between purchases and sales prices. Let

$$\delta^j(q, v) := \frac{\text{sgn}(q^j)v^j}{1 + v^j} \tag{1}$$

for each $j = 1, \dots, \ell$. Define a diagonal matrix $\Lambda(q, v)$ as a $\ell \times \ell$ matrix with diagonal elements $1 - \delta^j(q, v)$, $j = 1, \dots, \ell$. Then, the selling price vector associated with q is set to be given by $\Lambda(q, v)q$. Thus, spreads between buying and selling prices $q - \Lambda(q, v)q$ is equal to $(\delta^1(q, v)q^1, \dots, \delta^\ell(q, v)q^\ell)$. The spread factor $\delta^j(q, v)$ has the following property:

(i) the absolute value of buying price is greater or less than that of selling price depending upon whether it is positive or negative;
(ii) non-zero buying and selling prices are identical if and only if there are no trades in the markets, i.e., $v^j = 0$;
(iii) $|\delta^j(q, v)|$ is increasing in v^j and $0 \le |\delta^j(q, v)| < 1$.

It follows from (1) that for each $j = 1, \dots, \ell$ we have

$$\delta^j(q, v)q^j \ge 0\,; \tag{2}$$

$$\delta^j(q, v)q^j v^j > 0 \text{ if and only if } q^j v^j \ne 0. \tag{3}$$

Given a vector $q \in \mathbb{B}^\ell$ of buying prices, the following notation will be used to denote the vector of buying prices and reverse-signed selling prices:

$$q(v) := (q, -\Lambda(q, v)q) \in \mathbb{R}^\ell \times \mathbb{R}^\ell. \tag{4}$$

Now define

$$b^M := \frac{m}{M}\sum_{j=1}^{\ell} \bar{e}^j, \tag{5}$$

$$\Delta^M := [0, b^M] \subset \mathbb{R}_+. \tag{6}$$

$p^M \in \Delta^M$ will be the value of money. b^M is an upper bound for p^M. In our proof here buying prices are "normalized" to the unit closed ball, and selling prices and the value of money are expressed relative to buying prices.

2. We now define correspondences $\beta_i, \varphi_i, \mu_i$. Let $(x, v, q, p^M) \in X \times V \times \mathbb{B}^\ell \times \Delta^M$. For $i = 1, \dots, m$

$$\beta_i(x, v, q, p^M) := \left\{z_i \in X_i |\ q(v) \cdot z_i \le p^M M_i + 1 - \|q\|, z_i^C \ge 0\right\},$$
$$\beta_{m+1}(x, v, q, p^M) := \mathbb{B}^\ell.$$

For $i = 1, \dots, m$

$$\varphi_i(x, v, q, p^M) := \left\{z_i \in X_i \mid z_i^C \succ_i x_i^C\right\},$$
$$\varphi_{m+1}(x, v, q, p^M) := \left\{\bar{q} \in \mathbb{B}^\ell \mid \bar{q}(v) \cdot [\textstyle\sum_{i=1}^m x_i - (v, v)]\right.,$$
$$\left. > q(v) \cdot [\textstyle\sum_{i=1}^m x_i - (v, v)]\right\},$$
$$\mu_{m+2}(x, v, q, p^M) := \max\left\{\textstyle\sum_{i=1}^m x_i^B, \sum_{i=1}^m x_i^S\right\},$$

where the maximum is taken coordinatewise,

$$\mu_{m+3}(x, v, q, p^M) := \left\{\frac{q(v) \cdot (v, v)}{M}\right\}.$$

β_i and φ_i, $i = 1, \dots, m$, are budget and preference correspondences constrained to $X_i = [0, \bar{e}]$. The correspondence μ_{m+2} defines the *volume of trades*. μ_{m+3} sets the value of money so that the total value of money balance is exactly equal to the amount of money needed to pay for the difference between buying and selling values of commodities in carrying out total trades. φ_{m+1} sets the buying prices (and hence the selling prices) of commodities in such a way that the value of total excess demand for money in the economy is maximized. (Let us recall that $\bar{q}(v)$ in the definition of φ_{m+1} is defined as in the equation (4) with q replaced by $\bar{q}$. Keep in mind that this convention is used again when we write $q^*(v^*)$ and $q(v^*)$ in (10) that are defined as in (4) with q and/or v replaced by q^* and/or v^*.)

3. A slightly modified version of an abstract equilibrium existence theorem due to Shafer-Sonnenschein (1975) will be applied to the correspondences introduced above.

Given non-empty finite sets N_1 and N_2, a pair $\{(Z_i, \varphi_i, \beta_i)_{i \in N_1}, (Z_i, \mu_i)_{i \in N_2}\}$ composed of a family of ordered triples and a family of ordered pairs, where $\varphi_i : Z \to Z_i, \beta_i : Z \to Z_i$, and $\mu_i : Z \to Z_i$ are correspondences with $Z = \prod_{i \in N_1 \cup N_2} Z_i$, is called a *bisectoral social system.* An *equilibrium* for the bisectoral social system is an $x \in Z$ satisfying $x_i \in \beta_i(x)$, $\varphi_i(x) \cap \beta_i(x) = \emptyset$ for all $i \in N_1$, and $x_i \in \mu_i(x)$ for all $i \in N_2$.

Equilibrium Existence Lemma: *If, for every $i \in N_1 \cup N_2$, the set Z_i is a non-empty, compact, convex subset of a Euclidean space, φ_i is an open graph correspondence from Z to Z_i such that for every $z \in Z$, $z_i \notin \text{co}\,\varphi_i(z)$, and β_i and μ_i are non-empty and convex-valued correspondences from Z to Z_i where β_i's are continuous and μ_i's are upper hemi-continuous, then the bisectoral social system $\{(Z_i, \varphi_i, \beta_i)_{i \in N_1}, (Z_i, \mu_i)_{i \in N_2}\}$ has an equilibrium.*

The lemma follows from the proof of a theorem in Shafer-Sonnenschein (1975). We shall show later that $\beta_i, \varphi_i, \mu_i$ correspondences introduced in the previous step satisfy all the conditions of the above lemma. Thus, by applying the equilibrium existence lemma to the bisectoral social system defined by β_i, φ_i $(i = 1, \ldots, m+1)$ and μ_i $(i = m+2, m+3)$, one obtains $\xi^* = (x^*, v^*, q^*, p^{*M}) \in X \times V \times \mathbb{B}^\ell \times \Delta^M$ satisfying

$$q^* \in \mathbb{B}^\ell \tag{7}$$

$$q^*(v^*) \cdot x_i^* \le p^{*M} M_i + 1 - \|q^*\| \tag{8}$$

$$x_i^{*C} \ge 0 \quad \text{for} \quad i = 1, \ldots, m \tag{9}$$

$$q^*(v^*) \cdot [\textstyle\sum_{i=1}^m x_i^* - (v^*, v^*)] \ge q(v^*) \cdot [\textstyle\sum_{i=1}^m x_i^* - (v^*, v^*)] \tag{10}$$

$$\text{for all } q \in \mathbb{B}^\ell$$

$$\varphi_i(\xi^*) \cap \beta_i(\xi^*) = \emptyset \qquad \text{for } i = 1, \ldots, m+1 \tag{11}$$

$$v^* = \max\{\textstyle\sum_{i=1}^m x_i^{*B}, \sum_{i=1}^m x_i^{*S}\} \tag{12}$$

$$p^{*M} = \frac{q^*(v^*) \cdot (v^*, v^*)}{M} \tag{13}$$

4. Let us proceed to show the feasibility of the transaction allocation $x^* = (x_1^*, \ldots, x_m^*)$ and the nonzero property of commodity (buying) prices q^*.

We first show

$$\textstyle\sum_{i=1}^m x_i^{*B} = \sum_{i=1}^m x_i^{*S} = v^*. \tag{14}$$

If $\sum_i x_i^{*B} \neq \sum_i x_i^{*S}$, then we must have $(\sum_i x_i^{*B} - v^*) \neq 0$ or $(\sum_i x_i^{*S} - v^*) \neq 0$. It therefore follows from (10) that

$$q^*(v^*) \cdot [\sum_{i=1}^m x_i^* - (v^*, v^*)] > 0$$

and $\| q^* \| = 1$. Hence, one obtains from (8) and (13) that

$$q^*(v^*) \cdot [\sum_{i=1}^m x_i^* - (v^*, v^*)] \leq q^*(v^*) \cdot \sum_{i=1}^m x_i^* - p^{*M} M \leq 0$$

which contradicts the above strict inequality. This establishes (14).

We now show that in (8) one actually has

$$q^*(v^*) \cdot x_i^* = p^{*M} M_i + 1 - \|q^*\| \quad \text{for } i = 1, \ldots, m. \tag{15}$$

Recall that (14) implies

$$\sum_{i=1}^m x_i^{*C} = \sum_{i=1}^m e_i \tag{16}$$

which in turn implies

$$0 \leq x_i^{*C} \leq \sum_{i=1}^m e_i < \bar{e} \text{ for any } i = 1, \ldots, m. \tag{17}$$

Thus, if we had $q^*(v^*) \cdot x_i^* < p^{*M} M_i + 1 - \|q^*\|$ for some i, then by the local nonsatiation of preferences we would obtain $\varphi_i(\xi^*) \cap \beta_i(\xi^*) \neq \emptyset$, which contradicts (11).

It now follows from (13), (14), and (15) that

$$\| q^* \| = 1. \tag{18}$$

In addition to (18) one can prove

$$q^{*j} \neq 0 \quad \text{for every} \quad j = 1, \ldots, \ell. \tag{19}$$

Indeed, assume we had $q^{*h} = 0$ for some $1 \leq h \leq \ell$. It follows from (17) and the property of the possibility of individual utility enhancement for feasible allocations that for some agent i there is a consumption vector $y \in \mathbb{R}_+^\ell$ such that $y^j = x_i^{*Cj}$ for all $j \neq h$ and $y \succ_i x_i^{*C}$. Put $z^B = (y - e_i)^+$ and $z^S = (y - e_i)^-$ where for any $t \in \mathbb{R}^\ell$, $t^+ := \max\{t, 0\}$ and $t^- := \max\{-t, 0\}$ with the maximum taken coordinatewise. Write $z = (z^B, z^S)$. Then, we have $z^C = z^B - z^S + e_i = y \geq 0$, $z^{Bj} - z^{Sj} = x_i^{*Bj} - x_i^{*Sj}$ for all $j \neq h$, $z^B = (z^B - z^S)^+$, and $z^S = (z^B - z^S)^-$. To check z is budget feasible for i, define $J_B := \{j | \, x_i^{*Bj} > x_i^{*Sj}\}$ and $J_S := \{j \mid x_i^{*Bj} < x_i^{*Sj}\}$. Then, since $q^{*h} = 0$ we have

$$\begin{aligned} &q^*(v^*) \cdot x_i^* - q^*(v^*) \cdot z \\ &\quad = \sum_{j=1}^\ell \left(q^{*j} x_i^{*Bj} - \left(1 - \delta^j(q^*, v^*)\right) q^{*j} x_i^{*Sj}\right) - \sum_{j \in J_B} q^{*j} \\ &\quad \times \left(x_i^{*Bj} - x_i^{*Sj}\right) + \sum_{j \in J_S} \left(1 - \delta^j(q^*, v^*)\right) q^{*j} \left(x_i^{*Sj} - x_i^{*Bj}\right) \\ &\quad = \sum_{j \in J_B} \delta^j(q^*, v^*) q^{*j} x_i^{*Sj} + \sum_{j \in J_S} \delta^j(q^*, v^*) q^{*j} x_i^{*Bj} \geq 0 \end{aligned}$$

where the last inequality follows from the property of δ^j described by (2). This shows z is budget feasible for i. Hence, $z^C \succ_i x_i^{*C}$ contradicts (11). Thus (19) must be true.

5. We now show that

$$p^{*M} > 0\,. \tag{20}$$

If $p^{*M} = 0$, (13) implies

$$q^*(v^*) \cdot (v^*, v^*) = \textstyle\sum_{j=1}^{\ell} \delta^j(q^*, v^*) q^{*j} v^{*j} = 0\,.$$

It then follows from (2), (3), and (19) that $v^* = 0$ which implies by (12)

$$\begin{aligned} x_i^{*B} = x_i^{*S} &= 0 \text{ for all } i = 1, \ldots, m, \text{ and} \\ q^*(v^*) &= (q^*, -q^*)\,. \end{aligned}$$

Since the transaction allocation $(0, 0)$ is not Pareto optimal by the hypothesis of the theorem, there exists a feasible transaction allocation $(x_1, \ldots, x_m)$ that Pareto improves $(x_1^*, \ldots, x_m^*)$ so that $x_i^C \succ_i x_i^{*C}$ for all i and $\sum_{i=1}^m x_i^B = \sum_{i=1}^m x_i^S$. But (11) implies that for all $i = 1, \ldots, m$ we must have

$$q^* \cdot x_i^B - q^* \cdot x_i^S = q^*(v^*) \cdot x_i > p^{*M} M_i \geq 0\,.$$

One thus obtains $q^* \cdot (\sum_{i=1}^m x_i^B - \sum_{i=1}^m - x_i^S) > 0$ contradicting the feasibility of the transaction allocation $(x_1, \ldots, x_m)$. Therefore, we must have (20). This argument also establishes

$$v^* \neq 0. \tag{21}$$

6. Let $\bar{e}_k := k\bar{e}$ for each positive integer $k = 1, 2, \ldots$ and replace $\bar{e}$ by $\bar{e}_k$ in Steps 1–5. Then, the arguments in the previous steps establish the existence of $\xi_k^* = (x_k^*, v_k^*, q_k^*, p_k^{*M})$ satisfying the properties (7)–(21) where ξ^* is replaced by ξ_k^*. We show ξ_k^*, $k = 1, 2, \ldots$, are bounded. For this purpose we show that for every $i = 1, \ldots, m$ one has

$$(x_i^{*Bj})(x_i^{*Sj}) = 0 \quad \text{for all} \quad j = 1, \ldots, \ell \tag{22}$$

where the suffix k is omitted from now on in our present step of arguments. Put $y_i^B := (x_i^{*B} - x_i^{*S})^+$, $y_i^S := (x_i^{*B} - x_i^{*S})^-$, and $y_i := (y_i^B, y_i^S)$. Then, we have $0 \leq y_i \leq x_i^*$ and $y_i^C = x_i^{*C}$. Define sets, J_B and J_S, of indices of commodities as in Step 4. We prove the property (22) by showing $y_i = x_i^*$. Assume $y_i \neq x_i^*$. Then, we must have $x_i^{*Sj} > 0$ for some $j \in J_B$ or $x_i^{*Bj} > 0$ for some $j \in J_S$. It therefore follows from (2), (3), (19), and (21) that

$$
\begin{aligned}
&q^*(v^*)\cdot x_i^* - q^*(v^*)\cdot y_i \\
&= \textstyle\sum_{j=1}^{\ell} \left(q^{*j}(v^*) x_i^{*Bj} - \left(1-\delta^j(q^*,v^*)\right) q^{*j}(v^*) x_i^{*Sj} \right) \\
&\quad - \textstyle\sum_{j\in J_B} q^{*j}(v^*) \left(x_i^{*Bj} - x_i^{*Sj} \right) \\
&\quad + \textstyle\sum_{j\in J_S} \left(1-\delta^j(q^*,v^*)\right) q^{*j}(v^*) \left(x_i^{*Bj} - x_i^{*Sj} \right) \\
&\geq \textstyle\sum_{j\in J_B} \delta^j(q^*,v^*) q^{*j}(v^*) x_i^{*Sj} + \sum_{j\in J_S} \delta^j(q^*,v^*) q^{*j}(v^*) x_i^{*Bj} \\
&> 0\,.
\end{aligned}
$$

Thus, one has $q^*(v^*)\cdot y_i < p^{*M} M_i$. But since $y_i^C = x_i^{*C}$, by (17) it contradicts (11) and the local nonsatiation of preferences. This establishes (22).

Now, for every $i = 1,\ldots,m$, one has $x_i^{*B} - x_i^{*S} + e_i = x_i^{*C} \geq 0$. Thus, for each i

$$x_i^{*S} - x_i^{*B} \leq e_i \leq \textstyle\sum_{i=1}^{m} e_i\,. \tag{23}$$

In view of (22), (23) implies that

$$0 \leq x_i^{*S} \leq e_i \leq \textstyle\sum_{i=1}^{m} e_i \quad \text{for every } i = 1,\ldots,m\,. \tag{24}$$

By (14), (24) implies

$$0 \leq x_i^{*B} \leq \textstyle\sum_{i=1}^{m} x_i^{*B} = \sum_{i=1}^{m} x_i^{*S} \leq \sum_{i=1}^{m} e_i \quad \text{for every} \quad i = 1,\ldots,m\,. \tag{25}$$

Since (24) and (25) hold for all $k = 1,2,\ldots$, it follows from (13) and (14) that $\xi_k^* = (x_k^*, v_k^*, q_k^*, p_k^{*M})$, $k = 1,2,\ldots$, are bounded. Let us denote a limit point of the sequence ξ_k^* by $\xi^* = (x^*, v^*, q^*, p^{*M})$ again and assume without loss of generality that the sequence ξ_k^* itself converges to ξ^*. As every ξ_k^* satisfies (13), (14), (15) and (18), so does ξ^*. The argument in Step 4 to obtain (19) also applies to the limit point ξ^*, and hence so does the argument in Step 5 to obtain (20).

Replacing X_i by $X_i^* := \mathbb{R}_+^\ell \times \mathbb{R}_+^\ell$, define β_i^* and φ_i^* for each $i = 1,\ldots,m$ just as β_i and φ_i are defined in Step 2. Then, one must have

$$\varphi_i^*(\xi^*) \cap \beta_i^*(\xi^*) = \emptyset \quad \text{for} \quad i = 1,\ldots,m, \tag{26}$$

because, otherwise, one would have

$$\varphi_i(\xi_k^*) \cap \beta_i(\xi_k^*) \neq \emptyset$$

for k large enough in contradiction to (11) since each x_i on the budget hyperplane has local cheaper points as to be shown in the last step of our proof.

7. Define $p^* = (p^{*B}, -p^{*S})$ by

$$p^{*Bj} = \frac{q^{*j}}{p^{*M}} \quad \text{for} \quad j = 1, \ldots, \ell \quad \text{and}$$
$$p^{*Sj} = \frac{\left(1 - \delta^j(q^*, v^*)\right) q^{*j}}{p^{*M}} \quad \text{for} \quad j = 1, \ldots, \ell.$$

Then, by (14), (15), (18),(20), and (26), $((x_1^*, \ldots, x_m^*), p^*)$ is a monetary equilibrium. It remains to show that the correspondences $\beta_i, \varphi_i, \mu_i$ introduced in Step 2 satisfy the conditions of the Equilibrium Existence Lemma.

8. The nonempty-valuedness of β_{m+1} and μ_i, $i = m+2, m+3$, is trivial, and that of β_i, $i = 1, \ldots, m$, follows from $p^M \geq 0$, $q \in \beta^\ell$, and $0 \in X_i$. Note that in the lemma φ_i's are allowed to be empty-valued. The upper hemi-continuity of μ_i, $i = m+2, m+3$, and the continuity of β_{m+1} are immediate. That φ_i has an open graph is straightforward for $i = m+1$ and is a consequence of the continuity of preferences for $i = 1, \ldots, m$. The upper hemi-continuity of β_i, $i = 1, \ldots, m$, being straightforward, the only property one needs to show is the lower hemi-continuity of β_i, $i = 1, \ldots, m$. For this purpose, it is sufficient to prove existence of local cheaper points for any transaction vector lying on a budget hyperplane. We state it as a lemma.

Lemma: *Given $x_i \in \mathbb{R}_+^\ell \times \mathbb{R}_+^\ell$, $q \in \mathbb{B}^\ell$, $p^M \geq 0$, $v \in \mathbb{R}_+^\ell \times \mathbb{R}_+^\ell$ satisfying $q(v) \cdot x_i = p^M M_i + 1 - \|q\|$, there is a sequence $\{x_{in}\}_n$ in X_i such that $x_{in} \to x_i$ and $q(v) \cdot x_{in} < q(v) \cdot x_i$ for all n.*

The lemma is an immediate consequence of the following fact:

Fact: *Let $p\,(\neq 0) \in \mathbb{R}^n$ and $H(p) = \{z \in \mathbb{R}^n | \ p \cdot z = 0\}$. If one has $H(p) \cap \mathbb{R}_{++}^n = \emptyset$, then $p \in \mathbb{R}_+^n \cup \mathbb{R}_-^n$.*

Let $p \neq 0$ and $H(p) \cap \mathbb{R}_{++}^n = \emptyset$. Let $J_+ = \{j | \ p^j > 0\}$, $J_0 = \{j | \ p^j = 0\}$, and $J_- = \{j | \ p^j < 0\}$. If the conclusion of the fact were false, then one would have $p \notin \mathbb{R}_+^n \cup \mathbb{R}_-^n$. It would imply $J_+ \neq \emptyset$ and $J_- \neq \emptyset$. Let $x \in \mathbb{R}^n$ be defined by

$$x^j = \begin{cases} 1 & \text{for} \quad j \in J_0 \\ \frac{1}{p^j} & \text{for} \quad j \in J_+ \\ -\frac{\#J_+}{(\#J_-)p^j} & \text{for} \quad j \in J_- \end{cases}$$

where # indicates the number of elements in a set. Then, $x \in \mathbb{R}_{++}^n$ and $p \cdot x = 0$, i.e., $x \in H(p)$, contradicting $H(p) \cap \mathbb{R}_{++}^n = \emptyset$. This proves the above fact.

Let x_i, q, p^M, and v be given as in the lemma. Since $q(v) = (q, -\Lambda(q, v)q) \notin \mathbb{R}_+^{2\ell} \cup \mathbb{R}_-^{2\ell}$, by applying the above fact to the hyperplane $H\,(q(v)) \subset \mathbb{R}^{2\ell}$, one obtains $y_i \in \mathbb{R}_{++}^{2\ell}$ such that $q(v) \cdot y_i < 0$. By shortening the length of the vector y_i if necessary, one can assume w.l.o.g. $y_i \in X_i$. Now, for each $n = 1, 2, \ldots$

define $x_{in} := x_i + \frac{1}{n}(y_i - x_i) \in X_i$. Then, one has $q(v) \cdot x_{in} < p^M M_i + 1 - \|q\|$ and $x_{in} \to x_i$.

This establishes the lemma and completes the proof of the theorem. □

References

1. Bergstrom, T.C.: How to discard 'free disposability' – at no cost. Journal of Mathematical Economics **3**, 131-134 (1976)
2. Duffie, D.: Money in general equilibrium theory. In: Handbook of Monetary Economics, Vol. I (B.M. Friedman and F.H. Hahn eds.), pp.81-100 Elsevier Science Publishers 1990
3. Foley, D.: Economic equilibrium with costly marketing. Journal of Economic Theory **2**, 276-291 (1970)
4. Hahn, F.: On some problems of proving the existence of an equilibrium in a monetary economy. In: The Theory of Interest Rates (F. Hahn and F. Brechling eds.), pp.126-135 Macmillan, London 1965
5. Hahn, F.: Equilibrium with transaction costs. Econometrica **39**, 417-439 (1971)
6. Hayashi, T.: The non-Pareto efficiency of initial allocation of commodities and monetary equilibrium: An inside money economy. Journal of Economic Theory **7**, 173-187 (1974)
7. Heller, W.: The holding of money balances in general equilibrium. Journal of Economic Theory **7**, 93-108 (1974)
8. Kurz, M.: Arrow-Debreu equilibrium of an exchange economy with transaction cost. International Economic Review **15**, 699-717 (1974)
9. Kiyotaki, N., Wright, R.: On money as a medium of exchange. Journal of Political Economy **97**, 927-954 (1989)
10. Kurz, M.: Equilibrium with transaction cost and money in a single market exchange economy. Journal of Economic Theory **7**, 418-452 (1974b)
11. Lerner, A.: Money as a creature of the state. Proceedings of American Economic Association **37**, 312-317 (1947)
12. Okuno, M.: Essays on monetary equilibrium in a sequence of markets. Technical Report 120, Institute for Mathematical Studies in the Social Sciences, Stanford University 1973.
13. Ostroy, J., Starr, R.: Money and the decentralization of exchange. Econometrica **42**, 1093-1113 (1973)
14. Shafer, W., Sonnenschein, H.: Equilibrium in abstract economies without ordered preferences. Journal of Mathematical Economics **2**, 345-348 (1975)
15. Starr, R.: The structure of exchange in barter and monetary economies. Quarterly Journal of Economics **86**, 290-302 (1972)
16. Starr, R.: The price of money in a pure exchange monetary economy with taxation. Econometrica **42**, 45-54 (1974)
17. Starrett, D.: Inefficiency and the demand for 'money' in a sequence economy. Review of Economic Studies **40**, 437-448 (1973)
18. Yamazaki, A.: Monetary equilibria in a continuum economy with general transaction technologies. RUEE Working Paper 89-39, Hitotsubashi University 1989

19. Yamazaki, A.: Equilibrium in economies with incomplete markets and outside money: Transactions costs and existence. Department of Economics, Hitotsubashi University 1991

Subject Index

Instructions for Authors

A. General

1. Papers submitted for publication will be considered only if they have not been and will not be published elsewhere without permission from the publisher and the Research Center of Mathematical Economics.

2. Every submitted paper will be subject to review. The names of reviewers will not be disclosed to the authors or to anybody not involved in the editorial process.

3. The authors are asked to transfer the copyright to their articles to Springer-Verlag if and when these are accepted for publication.

The copyright covers the exclusive and unlimited rights to reproduce and distribute the article in any form of reproduction. It also covers translation rights for all languages and countries.

4. Manuscript must be written in English. One original and 3 sets of photocopies should be submitted to the following address.

Professor Toru Maruyama
Department of Economics
Keio University
2-15-45 Mita, Minato-ku, Tokyo
108-8345 Japan

5. **Offprints**: Twenty-five (25) **offprints** of each paper will be supplied free of charge. Additional offprints can be ordered at cost price.

B. Preparation of Manuscript

1. Manuscripts must be double-spaced (not 1.5), with wide margins (at least 25 mm), and large type (at least 12 point) on one side of A4 paper. Any manuscript that does not meet these requirements will be returned to the author immediately for retyping.

2. All manuscripts would finally be composed using our Springer LaTeX macro package. The Springer macros use standard LaTeX packages - if you do not have these packages refer to www.dante.de/cgi-bin/ctan-index. If authors make manuscripts by word-processing software other than TeX, please follow our styles as possible. For authors who prepare their manuscripts by TeX, we strongly recommend to visit our homepage, http://www.springer.de/economics/authors/interinf.html, where you can download all the necessary macro packages with instructions for LaTeX2e, LaTeX 2.09 and plain tex. For support, please contact `texhelp@springer.de`.

3. **The title page** must include: title; short (running) title of up to 60/85 characters; first names and surnames of all coauthors with superscript numerals indicating their affiliations; full street addresses of all affiliations; address to which proofs should be sent; fax number; and any footnotes referring to the title (indicated by asterisks*).

4. **Summary/abstract**: Each paper must be preceded by a summary/an abstract, which should not exceed 100 words.

5. **The Journal of Economic Literature index number (JEL classification)** should be indicated and the statement of the **2000 Mathematics Subject Classification (MSC) numbers** is desirable. You can check JEL classification with Internet at `http://ideas.uqam.ca/ideas/data/JEL` as well as 2000 MSC numbers at `http://e-math.ams.org/msc`.

6. **Main text**: All tables and figures must be cited in the text and numbered consecutively with Arabic numerals ac-

cording to the sequence in which they are cited. Please mark the desired position of tables and figures in the left margin of the manuscript.

Do not italicize, underscore, or use boldface for headings and subheadings. Words that are to be italicized should be underscored in pencil.

Abbreviations must be spelled out at first mention in summary/abstract and main text. Abbreviations should also be spelled out at first mention in figure legends and table footnotes.

Short equations can be run in with the text. Equations that are displayed on a separate line should be numbered.

7. **References**: The list of references should be in alphabetical order and include the names and initials of all authors (see examples below). Whenever possible, please update all references to papers accepted for publication, preprints or technical reports, giving the exact name of the journal, as well as the volume, first and last page numbers and year, if the article has already been published or accepted for publication.

When styling the references, the following examples should be observed:

Journal article:
1. or [F-M] Freed, D.S., Melrose, R.B.: A mod *k* index theorem. Invent. Math. **107**, 283-299 (1992)

Complete book:
2. or [C-S] Conway, J.H., Sloane, N.J.: Sphere packings, lattices, and groups (Grundlehren Math. Wiss. Bd. 290) Berlin Heidelberg New York: Springer 1988

Single contribution in a book:
3. or [B] Border, K.C.: Functional analytic tools for expected utility theory. In: Aliprantis, C.D. et al. (eds.): Positive operators, Riesz spaces and economics. Berlin Heidelberg New York: Springer 1991, pp. 69-88

8. **Citations in the text** should be either by numbers in square brackets, e.g. [1], referring to an alphabetically ordered and numbered list, or by the author's initials in square brackets, e.g. [F-M], or by author and year in parentheses, e.g. Freed and Melrose (1992). Any of these styles is acceptable if used consistently throughout the paper. In the third system, if a work with more than two authors is cited, only the first author's name plus "et al." need be given and if there is more than one reference by the same author or team of authors in the same year, then a, b, c should be added after the year both in the text and in the list of references.

9. **Tables** are to be numbered separately from the illustrations. Each table should have a brief and informative title. All abbreviations used in a table must be defined in a table footnote on first use, even if already defined in the text. In subsequent tables abbreviations need not be redefined. Individual footnotes should be indicated by superscript lowercase a, b, c, etc. Permission forms must be provided for any tables from previously published sources (same procedure as with figures; see below).

10. **Figures**: If you have access to suitable software, you can design your figures directly on a computer, but creation by other means is of course also possible. In any case the originals should be pasted into the camera- ready copy at the appropriate places and with the correct orientation. If necessary, the figures may be drawn overscale, but in this case suitably reduced copies should be pasted into the script.

If a figure has been published previously, acknowledge its source and submit written permission signed by author and publisher. The source must be included in the reference list. If a permission fee is required, it must be paid by the author. Responsibility for meeting this requirement lies entirely with the author.